X-treme Review

Math 6
Test Preparation

Authors
Celestine Marie Milanese and **Carolyn Peck**

Editors
Wayne Garnsey and **Paul Stich**

N&N Publishing Company, Inc.
18 Montgomery Street, Middletown, New York 10940-5116

Ordering and Information
1-800-NN 4 TEXT

Internet: www.nn4text.com   email: nn4text@nandnpublishing.com

DEDICATION

This book is dedicated in loving memory of Michael J. Scoba Sr. and Carolina Gonzalez. We know you are both very proud of our accomplishments. We love you and miss you!

ACKNOWLEDGEMENTS

The authors would like to thank the N&N production crew for their patience and talent in converting our "symbols," and Wayne Garnsey and Paul Stich for their unlimited support.

Thanks to Nana and "El Gitano Discreto" for your encouraging phone calls and e-mails, you are of our biggest fans! To our families thank you for all of your help and support.

And last but not definitely not least, many thanks to our proofreaders and problem solvers.

We thank all of you from the bottom of our hearts.

Front Cover: Photo courtesy of Image100 Ltd 2001©

CHELLIE SCOBA MILANESE & CAROLYN GONZALEZ PECK

Teachers: Wappingers Junior High School & Roy C. Ketcham High School
              Wappingers Falls, NY

© Copyright 2007

N&N Publishing Company, Inc.

Internet: www.nn4text.com   phone: 1-800-NN 4 TEXT   email: nn4text@nandnpublishing.com

SAN # - 216-4221   ISBN # - 0935487-84-0
1 2 3 4 5 6 7 8 9 10 BookMart Press 2013 2012 2011 2010 2009 2008 2007

TABLE OF CONTENTS

To The Student

Hello Sixth Grader! This *X-treme Math 6 Review* is designed to help you achieve success on the New York State 6th Grade Math Assessment Test.

Each chapter contains:

- ## VOCABULARY
  These words and phrases help you identify ideas and operations in Math and are often used as correct responses to questions throughout the lesson.

- ## SPECIFIC TOPICS
  Specific Topics are the "Strands", also called the Key (Main) Ideas. They are to be learned in preparation for the 6th Grade Math Test.

- ## PRACTICE
  Each Lesson has many practice questions which are the "Bands," also called the Performance Indicators. You must be able to understand and do for success on the 6th Grade Math Test.

- ## X-TREME NOTES, HINTS, AND REMEMBER
  These special "Notes" and "Remember" hints give you pointers and mental aids to remember the main ideas used on the 6th Grade Math Test.

- ## TEST PREPS
  These questions are similar to the questions you will have to answer correctly on the 6th Grade Math Test – two practice tests are found at the end of this book.

- ## TWO COMPLETE TESTS
  These practice tests give you the opportunity to rehearse with questions on the level of the Assessment Test in order to do well on the 6th Grade Math Test.

- ## SYMBOLS AND REFERENCES
  These are a guide to many of the standard signs and symbols used in math operations (page 123-124).

Make Your Study X-treme!
Good luck!

LESSON ONE

NUMBER SYSTEMS

Vocabulary

These words and phrases are associated with Number Systems and may be used when answering questions in this chapter. Definitions can be found in the Glossary/Index at the back of this *X-treme Review*.

absolute value	identity element of addition
additive inverse	identity element of multiplication
associative property	rational number
common denominators	repeating decimal
commutative property	standard form
counting numbers	terminating decimal
distributive property	whole number
equivalent fractions	zero property of addition
fraction	zero property of multiplication

The Native American Number System

Long before Columbus and other Europeans came to the Americas, the ancient Aztecs and Maya developed advanced number systems. They used base-20 for their systems. The number system most used throughout the world today is a base-10 system. In recent decades anthropologists have learned a great deal about Native American mathematics from archaeological excavations and from people who still use mathematical methods developed by their ancestors. These concepts addressed the need to express large numbers as well as insight into the current base-10 system that is so important in our own culture.

0	1	2	3	4
5	6	7	8	9
10	11	12	13	14
15	16	17	18	19
20	21	22	23	24
25	26	27	28	29

Mayan positional number system

At the left is the 20-month Aztec Calendar Stone (base-20), one of the most famous symbols of Mexico.

Historically, the Aztec name for the huge basaltic monolith is Cuauhxicalli Eagle Bowl, but it is universally known as the Aztec Calendar or Sun Stone. It was during the reign of the 6th Aztec monarch in 1479 that this stone was carved and dedicated to the principal Aztec deity: the Sun. The stone has both mythological and astronomical significance. It weighs almost 25 tons, has a diameter of just under 12 feet, and a thickness of 3 feet.

On December 17th, 1790 the stone was discovered, buried in the "Zocalo" (the main square) of Mexico City. The viceroy of New Spain at the time was don Joaquin de Monserrat, Marquis of Cruillas. Afterwards it was embedded in the wall of the Western tower of the metropolitan Cathedral, where it remained until 1885. At that time, it was transferred to the national Museum of Archaeology and History by order of the then President of the Republic, General Porfirio Diaz.

http://www.mayacalendar.com MAYA WORLD STUDIES CENTER, MERIDA, YUCATAN, MEXICO

LESSON 1: NUMBER SYSTEMS

1.1 WHOLE NUMBERS

(1) _____ begin at 1 and go on forever in a positive direction.

Example: 1, 2, 3, 4, …

(2) _____ begin at 0 and go on forever in a positive direction.

Example: 0, 1, 2, 3, 4, …begin at 0 and go on forever in a positive direction.

Use a place value chart to help read whole numbers.

Trillions	Hundred-billions	Ten-billions	Billions	Hundred-millions	Ten-millions	Millions	Hundred-Thousands	Ten-Thousands	Thousands	Hundreds	Tens	Ones
5	0	6	4	3	2	0	1	9	7	8	5	2

The (3) _____ of a number is when a number is written with each digit in a place value.

> The standard form of the above number: **5,064,320,197,852**   This number is read, "Five trillion, sixty-four billion, three hundred twenty million, one hundred ninety-seven thousand, eight hundred fifty-two."

Remember: A comma is placed after every third number starting from the ones place *X-treme*   value.

PRACTICE

Directions: Write each number in standard form.

1   six million, two-hundred thousand, four hundred eight-five

_____

2   seventy trillion, eight billion, twenty-nine thousand, six

_____

3   forty-five billion, ninety-six million, thirty-thousand

_____

4   two-hundred fifty trillion, nine million, one-hundred seven

_____

Directions: Write the place value of the underlined digit.

5   8,613,542,304   _____

6   69,843,210,473   _____

7   903,864,501,233,100   _____

Directions: Write each number in words.

8   72,184,000,000

_____

9   7,230,702,300,000

_____

10   45,000,250,175

_____

1.2 PROPERTIES FOR ADDITION AND MULTIPLICATION

The (1) _____ states that the sum (addition) or product (multiplication) of any two numbers is unchanged by the order of those numbers.

Example: The order of the numbers for a and b can be reversed with the same result:

$$a + b = b + a \qquad\qquad a \cdot b = b \cdot a$$

$$3 + 4 = 4 + 3 \qquad\qquad 2 \cdot 5 = 5 \cdot 2$$

The (2) _____ states that the sum or product of a set of numbers is the same, regardless of how the numbers are ordered (grouped).

Example: The order for a, b, and c can be changed with the same result:

$$a + (b + c) = (a + b) + c \qquad\qquad a \cdot (b \cdot c) = (a \cdot b) \cdot c$$

$$8 + (5 + 4) = (8 + 5) + 4 \qquad\qquad 3 \cdot (6 \cdot 7) = (3 \cdot 6) \cdot 7$$

The (3) _____+_____ states that the product of the sum or difference of a set of numbers is the same as the sum or difference of their products.

Example: Multiplying the sum of two numbers by a third number gives the same result as multiplying each of the two numbers independently by the third number:

$$a(b + c) = a \cdot b + a \cdot c$$

$$3(6 + 4) = 3 \cdot 6 + 3 \cdot 4$$

The (4) _____ is the number which, when added to any number, yields the given number.

Remember: The identity element for addition is zero, because $a + 0 = a$ and $0 + a = a$.
X-treme

The (5) _____ is the number which, when multiplied by any number, yields the given number.

Remember: The identity element for multiplication is one because $a \cdot 1 = a$ and $1 \cdot a = a$.
X-treme

The (6) _____ states that the result of two real numbers when combined will result in the identity element.

Remember: When a number is added to its additive inverse, the sum is always zero.
X-treme

Example 1: 8 + -8 = 0

When a number is multiplied by its multiplicative inverse, the product is always one.

Example 2: $7 \times \dfrac{1}{7} = 1$

The (7) _____ states that the sum of a number and zero is that same number.

Example 3: a + 0 = a

The (8) _____ states that the product of any number and zero is always zero.

Example 4: a · 0 = 0

PRACTICE

Directions: Use the properties to find each missing number. Then, on the longer blank at the right, name the property used.

1 15 + 0 = _____   _____

2 21 · 1 = _____   _____

3 13 + 10 = _____ + 13   _____

4 4 · (7 + 5) = 4 · _____ + _____ · 5   _____

5 9 · (3 · 2) = (9 · _____) · 2   _____

6 8 · 12 = _____ · 8   _____

7 16 · 0 = _____   _____

1.3 RATIONAL NUMBERS

A (1) _____ is any number that can be written as a ratio $\frac{a}{b}$, where a and b are integers and $b \neq 0$.

Example 1: 0.5 $\frac{2}{3}$ 4

Rational numbers can be either terminating decimals or repeating decimals.

A (2) _____ has a fixed number of digits.

Example 2: $\frac{5}{8}$ = 0.625 $\frac{1}{4}$ = 0.25 $\frac{3}{5}$ = 0.6

A (3) _____ has one digit that repeats or a group of digits that repeat.

Example 3: $\frac{1}{3}$ = $0.\overline{33}...$ $\frac{5}{6}$ = $0.8\overline{3}...$ $\frac{2}{7}$ = $0.\overline{285714}...$

Remember: When changing a fraction to a decimal, divide the numerator by the denominator.

Example 4: $\frac{5}{8}$ or   5 ÷ 8   or $8\overline{)5.000}$ with quotient .625

Remember: When changing a decimal to a fraction, think of the fractional equivalents of place values. In other words, "read" the decimal correctly to write the fraction.

Example: To write 0.7 as a fraction, think: "seven tenths."

$$0.7 = \frac{7}{10}$$

Remember: Repeating decimals can also be written as fractions.

Use this pattern: $0.\overline{1} = \frac{1}{9}$ $0.\overline{01} = \frac{1}{99}$ $0.\overline{001} = \frac{1}{999}$

Example:

$$0.\overline{8} = 8 \cdot 0.\overline{1}$$
$$= 8 \cdot \frac{1}{9}$$
$$= \frac{8}{9}$$

$$0.\overline{82} = 82 \cdot 0.\overline{01}$$
$$= 82 \cdot \frac{1}{99}$$
$$= \frac{82}{99}$$

$$0.\overline{824} = 824 \cdot 0.\overline{001}$$
$$= 824 \cdot \frac{1}{999}$$
$$= \frac{824}{999}$$

Remember: When you compare and order rational numbers, it may be helpful to write each number as a decimal or to change each to an equivalent fraction with a common denominator.

Problem: Write the following numbers from least to greatest: $\frac{1}{3}, \frac{1}{4}, \frac{5}{6}, \frac{7}{12}$

Solution 1: (a) Change the fractions to decimals:

$$\frac{1}{3} = 0.\overline{3}$$

$$\frac{1}{4} = 0.25$$

$$\frac{5}{6} = 0.8\overline{3}$$

$$\frac{7}{12} = 0.58\overline{3}$$

(b) Compare the decimals: $0.25 < 0.3 < 0.583 < 0.83$

(c) Order the original fractions from least to greatest: $\frac{1}{4}, \frac{1}{3}, \frac{7}{12}, \frac{5}{6}$

Solution 2: (a) Change each fraction to an equivalent fraction with a common denominator. Then compare the numerators of each new fraction:

$$\frac{1}{3} = \frac{20}{60}$$

$$\frac{1}{4} = \frac{15}{60}$$

$$\frac{5}{6} = \frac{50}{60}$$

$$\frac{7}{12} = \frac{35}{60}$$

(b) Order the new fractions from least to greatest:

$$\frac{15}{60} < \frac{20}{60} < \frac{35}{60} < \frac{50}{60}$$

(c) Order the original fractions from least to greatest: $\frac{1}{4}, \frac{1}{3}, \frac{7}{12}, \frac{5}{6}$

PRACTICE

Directions: Write each fraction as a decimal.

1 $\frac{3}{4}$ = _____   2 $\frac{2}{5}$ = _____   3 $\frac{4}{9}$ = _____   4 $\frac{1}{6}$ = _____

Directions: Write each decimal as a fraction in lowest terms.

5 0.8 = _____   6 0.27 = _____   7 $0.\bar{6}$ = _____   8 0.15 = _____

Directions: Order the following rational numbers from least to greatest.

9 $\frac{1}{4}$, 0.3, $\frac{3}{5}$, $\frac{1}{3}$ = _____

10 $\frac{4}{7}$, 0.55, $\frac{1}{2}$, $\frac{2}{3}$ = _____

1.4 RATIONAL NUMBERS ON A NUMBER LINE

As was previously reviewed, the top number of a fraction (above the line), or the rational number, is known as the numerator, and the bottom number is known as the denominator.

On a number line, the denominator tells us the number of equal parts into which the number line is divided. The numerator tells us how many equal parts are being used.

There is a point on the number line corresponding to each fraction.

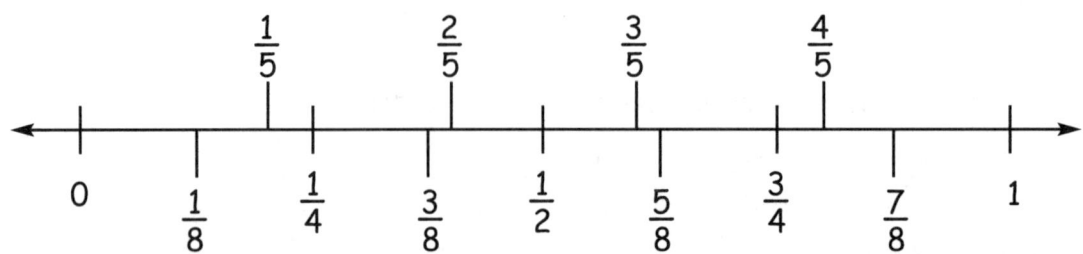

The (1) _____ of a number is its distance from 0 on the number line. | | is the symbol for absolute value.

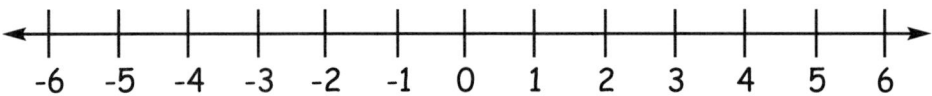

Example: Since |-4 | is 4 units from 0, the absolute value of -4 is 4. The distance from 0 to 5 is 5; therefore, the absolute value of |5 | is 5.

PRACTICE

Directions: Find the absolute value of each.

1 |6 | = ____   2 |-7 | = ____   3 $|\frac{2}{3}|$ = ____   4 $|-\frac{6}{11}|$ = ____

Place the following fractions on the number line provided.

5   $\frac{1}{3}, \frac{2}{3}$

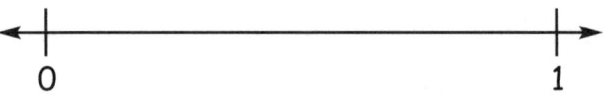

6   $\frac{1}{6}, \frac{5}{6}$

TEST PREP

Directions: Read the question or statement carefully. Then, choose the best response (circle the letter).

1   What is the value of the 2 in 18,205,694?
   A   20,000000
   B   2,000,000
   C   200,000
   D   20,000

2   Round 62,971 to the nearest thousand.
   F   60,000
   G   62,000
   H   62,900
   J   63,000

3   In what place is the 6 in 163,924,107?
   A   hundred millions
   B   millions
   C   ten millions
   D   ten thousands

4   What is 9,012,000 in word form?
   F   nine million twelve
   G   nine million, twelve thousand
   H   ninety million, twelve
   J   ninety million, twelve thousand

5   Write one million, two-hundred fifty-six thousand, ten in standard form.
   A   1,256,010
   B   1,256,100
   C   100,256,010
   D   100,256,100

6   Find the |-17|.
   F   -17
   G   71
   H   17
   J   -71

7   Which is not a rational number?
   A   4
   B $\frac{1}{2}$
   C   3.1415926535…
   D $0.\bar{3}$

8   Which shows the decimal for $\frac{8}{33}$?
   F   0.24
   G $2.\bar{4}$
   H   2.4
   J $0.\overline{24}$

9   Which set of fractions is in order from least to greatest?

   A $\frac{3}{4}, \frac{2}{3}, \frac{5}{7}$

   B $\frac{4}{7}, \frac{1}{2}, \frac{3}{5}$

   C $\frac{1}{2}, \frac{7}{8}, \frac{3}{4}$

   D $\frac{2}{3}, \frac{3}{4}, \frac{5}{6}$

10 What is the |30|?
   F   -30
   G   -3
   H   30
   J   3

11 Which of the following numbers are larger than $\frac{5}{8}$ and smaller than 0.98?

Part A
Circle all that apply.

0.75 $\frac{3}{5}$ 0.63 $\frac{8}{5}$

Part B
Explain in words why you circled those numbers.

_____

_____

_____

12 Given the following four numbers:

9   2.36908… $\frac{2}{3}$ $0.\overline{214}$

Part A
List all the numbers that are rational.

_____

Part B
Explain how you arrived at your answer.

_____

_____

_____

13 Place a dot on the number line to show the approximate position of each fraction. Label each dot with its number.

$\frac{1}{7} , \frac{2}{5} , \frac{1}{4} , \frac{2}{3} , \frac{9}{10}$

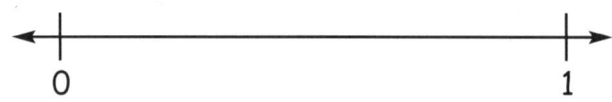

0   1

Vocabulary

These words and phrases are associated with Ratios, Rates, Proportions, and Percents and may be used when answering questions in this chapter. Definitions can be found in the Glossary/Index at the back of this *X-treme Review*.

base (of a percent)	equivalent ratios	proportional reasoning
congruent sides	extremes (of a proportion)	rate
congruent figures	means (of a proportion)	rate of interest
corresponding sides	percent	ratio
equivalent fractions	proportion	similar triangles

So, what is the real price?

One of the most common places where you will see percents used in the "real world," is shopping. Unfortunately, the percent you usually fail to remember (to add to the price) is the sales tax. Sales tax is added to the total of your purchase. The tricky thing about sales tax is that it changes as you travel state to state and within the state, county to county, and even sometimes from town to town.

In this lesson you will review procedures used to form ratios and rates, as well as proportions. You will then review using proportions to solve percent problems.

Source: http://ny.rand.org/stats/govtfin/salestax.html
Sales Tax Calculator: http://www.earthodyssey.com/sales_tax.html

Examples of NYS Sales Tax Rates by County   [Note: These Rates Change Frequently. 2005]

Albany 8.50	Essex 7.75	Onondaga 8.00	Schoharie 8.00	Mount Vernon ...8.38
Allegany 8.50	Franklin 8.00	*Nassau 8.62	Schenectady ...8.00	New Rochelle ...8.38
Broome 8.00	Fulton 8.00	*Orange 8.12	Seneca 8.00	White Plains ...7.88
Cattaraugus 8.00	Genesee 8.00	*New York City .8.38	Schuyler 8.00	Yonkers 8.38
Cayuga 8.00	Greene 8.00	Oswego 8.00	Steuben 8.00	Wyoming 8.00
Chautauqua 8.25	Hamilton 7.00	Oneida 9.50	Suffolk 8.62	Yates 8.00
Chemung 8.00	Herkimer 8.00	*Putnam 7.88	Sullivan 7.50	
Chenango 8.00	Jefferson 7.75	Ontario 7.00	Tioga 8.00	
Clinton 7.75	Livingston 8.00	Rensselaer 8.00	Tompkins 8.00	
Columbia 8.00	Lewis 7.75	Orleans 8.00	Ulster 8.00	
Cortland 8.00	Montgomery ...8.00	*Rockland 8.00	Warren 7.00	*Rates in these jurisdictions include 3/8% imposed for the benefit of the Metropolitan Commuter Transportation Dist.
Delaware 8.00	Madison 8.00	Otsego 8.00	Washington 7.00	
Dutchess 8.12	Niagara 8.00	Saratoga 7.00	Wayne 8.00	
Erie 8.75	Monroe 8.00	St. Lawrence ...7.00	Westchester ...7.38	

LESSON 2: RATIOS, RATES, PROPORTIONS, AND PERCENT

2.1 RATIOS AND RATES

A (1) _____ is a comparison of two numbers or two like quantities by division.

Remember: Ratios can be written in three different forms.

$$2 \text{ to } 3 \qquad \frac{2}{3} \qquad 2:3$$

Example: The ratio of footballs to soccer balls below is six to five.

$$6 \text{ to } 5 \qquad \frac{6}{5} \qquad 6:5$$

A (2) _____ is a ratio that compares quantities of different units.

Example: 5 boys to 12 girls $\dfrac{25 \text{ miles}}{1 \text{ hour}}$ $\dfrac{\$8.00}{2 \text{ hours}}$

Remember: To identify a rate, look for different units.

Example 1: Is the following ratio a rate? $\dfrac{3 \text{ softballs}}{5 \text{ softballs}}$

No, this is not a rate, because the unit is the same.

Example 2: 12 packages per carton

Yes, this is a rate, because it is a comparison between two different units (packages and cartons).

Example 3: 7 boys to 3 girls
Yes, because it is a comparison between two different units (boys and girls).

PRACTICE

Directions: Determine if the following ratios are rates? Write yes or no and explain why.

1   42 inches to 7 feet   _____

2    $\dfrac{5\ \text{teaspoons}}{3\ \text{teaspoons}}$ _____

3   25 miles to 17 miles   _____

4    $\dfrac{3\ \text{books}}{\$7.00}$ _____

5    $\dfrac{2\ \text{tables}}{36\ \text{chairs}}$ _____

2.2 EQUIVALENT RATIOS

Two ratios that are equal are called (1) _____.

Example: 8 to 12 is equivalent to 2 to 3

$\dfrac{3}{5}$ is equivalent to $\dfrac{6}{10}$

3 : 7 is equivalent to 30 : 70

Remember: To find equivalent ratios, multiply or divide both quantities of the ratio by the same amount. When the ratio is in fraction form, it is easier to find the equivalent ratios.

Example: Find two ratios equivalent to $\dfrac{8}{10}$

Multiply the numerator and denominator by 2:

$$\dfrac{8\ \cdot\ 2}{10\ \cdot\ 2}\ \begin{matrix}=\\=\end{matrix}\ \dfrac{16}{20}$$

Divide the numerator and denominator by 2:

$$\dfrac{8 \div 2}{10 \div 2}\ \begin{matrix}=\\=\end{matrix}\ \dfrac{4}{5}$$

PRACTICE

Directions: Give two ratios that are equivalent to the given ratio.

1   3 : 7   _____   _____

2   $\dfrac{9}{24}$ _____   _____

3   5 to 30   _____   _____

4   $\dfrac{8}{12}$ _____   _____

5   1 : 2   _____   _____

6   4 to 20   _____   _____

2.3 PROPORTIONS

A (1) _____ is an equation which states that two ratios are equivalent.

Example: $\dfrac{1}{2} = \dfrac{5}{10}$ $\dfrac{\$6.00}{books} = \dfrac{\$12.00}{2\ books}$

The (2) _____ of a proportion are the two middle terms in the ratios of a proportion.

Example: $\dfrac{2}{14} = \dfrac{4}{28}$ The means of this proportion are 14 and 4, the two middle terms.

The (3) _____ of a proportion are the first and last terms in the ratios of a proportion.

Example: $\dfrac{2}{14} = \dfrac{4}{28}$ The extremes of this proportion are 2 and 28, the two end terms.

Remember: Use cross multiplication to check proportionality of the two ratios.

X-treme

Example 1: Are the following two ratios proportional? Write yes or no and explain why.

$$2 \cdot 22 = 44 \qquad\qquad 9 \cdot 5 = 45$$

means ⟶ $\dfrac{2}{9} = \dfrac{5}{22}$ ⟵ extremes

<u>No, it is not a proportion, because the cross products are not equal.</u>

Example 2:

$$15 \cdot 2 = 30 \qquad\qquad 10 \cdot 3 = 30$$

$$\frac{15}{10} = \frac{3}{2}$$

← means

← extremes

Yes, it is a proportion, because the cross products are equal.

Example 3:

$$2 \cdot 5 = 10 \qquad\qquad 1 \cdot 10 = 10$$

extremes → $\dfrac{2 \text{ pies}}{10 \text{ cakes}} = \dfrac{1 \text{ pie}}{5 \text{ cakes}}$ ← means

Yes, it is a proportion, because the cross products are equal.

PRACTICE

Directions: Are the following two ratios proportional? Write yes or no and explain why.

1. $\dfrac{25 \text{ chairs}}{5 \text{ tables}} \overset{?}{=} \dfrac{5 \text{ chairs}}{1 \text{ table}}$

_____ _____

_____

2. $\dfrac{10}{9} \overset{?}{=} \dfrac{16}{14}$

_____ _____

_____

3. $\dfrac{3}{24} \overset{?}{=} \dfrac{4}{32}$

_____ _____

_____

4. $\dfrac{25}{5} \overset{?}{=} \dfrac{10}{2}$

_____ _____

_____

5. $\dfrac{5 \text{ minutes}}{15 \text{ gallons}} \overset{?}{=} \dfrac{2 \text{ minutes}}{6 \text{ gallons}}$

_____ _____

_____

6. $\dfrac{36}{28} \overset{?}{=} \dfrac{18}{14}$

_____ _____

_____

2.4 SIMILAR FIGURES

(1) _____ are triangles that have the same shape but not necessarily the same size; corresponding sides are in proportion and corresponding angles are congruent.

Example:

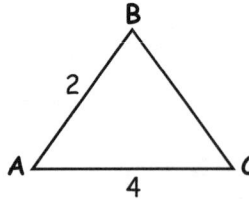 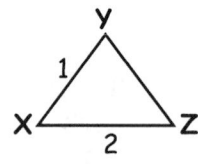

$$\frac{AB}{XY} = \frac{AC}{XZ}$$

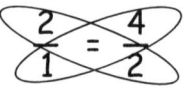

Remember: The symbol for similar is ~. The symbol for congruent is ≅.

Example: △A is similar to △B should be written: △A ~ △B

△C is congruent to △D should be written: △C ≅ △D

(2) _____ are sides in the same relative positions on two congruent or similar figures.

Example:

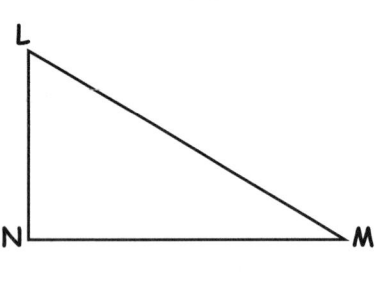 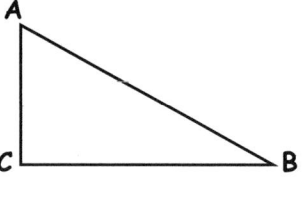

△LMN ~ △ABC   LM ~ AB   MN ~ BC   LN ~ AC

Remember: When looking for corresponding sides in two similar triangles, make sure your shapes are facing the same direction.

Example: △D ~ △E

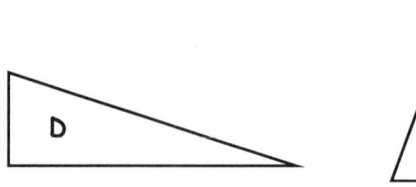 

Before you can compare corresponding sides, you need to rotate the triangles so that they face the same direction.

or

PRACTICE

Directions: Name the corresponding sides for each of the two similar triangles below.

1   △QRS ~ △JKL

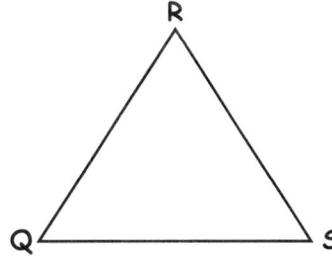   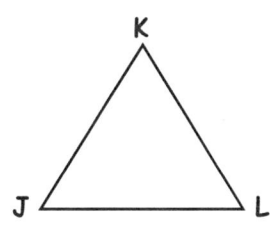

_____ ~ _____
_____ ~ _____
_____ ~ _____

2   △XYZ ~ △EFG

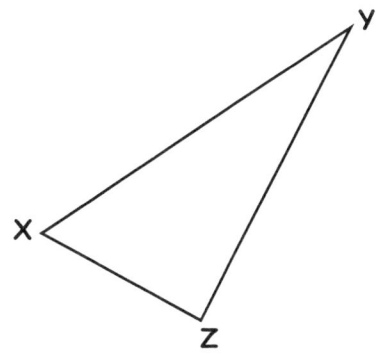   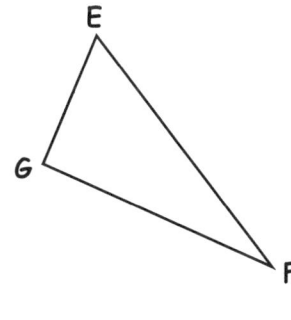

_____ ~ _____
_____ ~ _____
_____ ~ _____

(1) _____ is using the concept of proportions when analyzing and solving a mathematical situation.

Example: Use proportional reasoning to find the missing lengths in the following similar triangles. [not drawn to scale]

$\triangle EFG \sim \triangle RST$

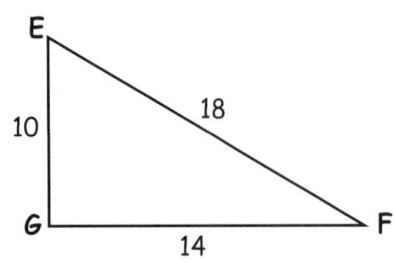 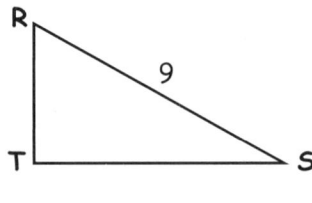

$$\frac{EF}{RS} = \frac{GF}{TS}$$

$$\frac{18}{9} = \frac{14}{x}$$

$$\frac{126}{18} = \frac{18x}{18}$$

$$7 = x$$

$$TS = 7$$

$$\frac{EF}{RS} = \frac{EG}{RT}$$

$$\frac{18}{9} = \frac{10}{x}$$

$$\frac{90}{18} = \frac{18x}{18}$$

$$5 = x$$

$$RT = 5$$

PRACTICE

Directions: Use proportional reasoning to find the missing lengths in the following similar triangles. [not drawn to scale]

1   Solve the triangles. [not drawn to scale]

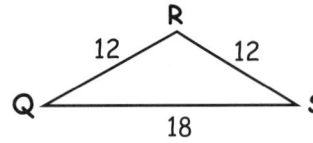 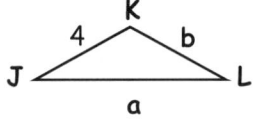

a   _____

b   _____

2 Solve the triangles. [not drawn to scale]

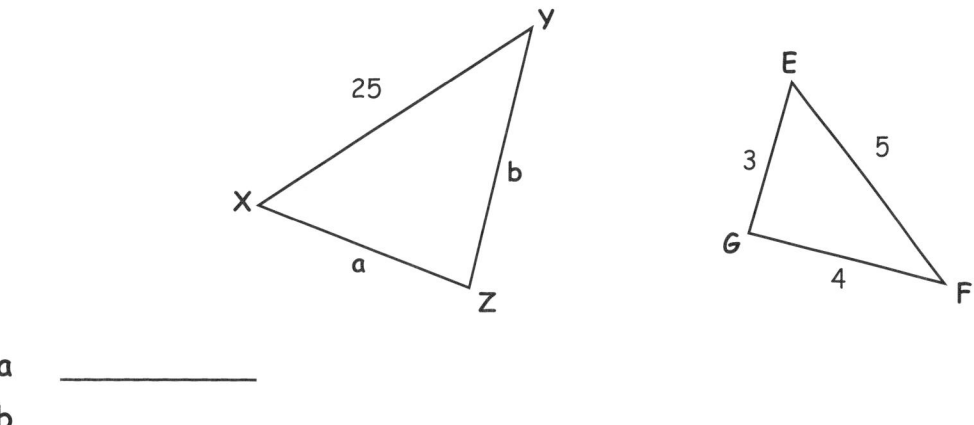

a _____

b _____

(1) _____ are two or more figures having exactly the same shape and size.

Directions: State whether the following polygons are similar, congruent, or neither.
Example 1:

Figure A ≅ B (congruent), because they are the same shape and size.

Example 2:

Figure C ~ D (similar), because they are the same shape, but different sizes.

Example 3:

Figures E and F are *neither* similar *nor* congruent, because they are not the same shape or size.

PRACTICE

Directions: State whether the following polygons are similar, congruent, or neither.

1   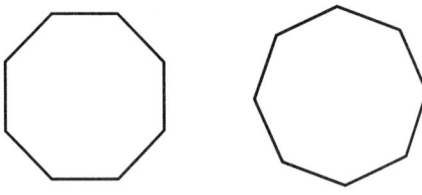   _____

2   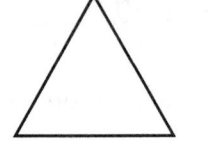 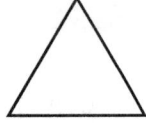   _____

3   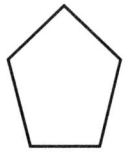 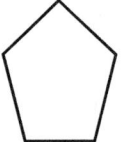   _____

4   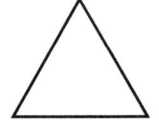 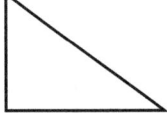   _____

5   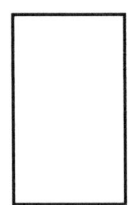 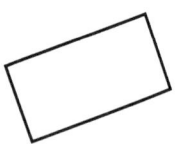   _____

6   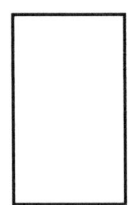 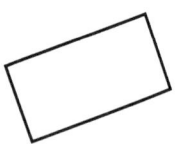   _____

2.5 PERCENT

A (1) _____ is one part in a hundred.

Example: $0.50 is fifty cents out of $1.00 or 100 cents.

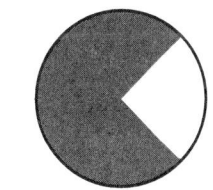 **Remember:** Percents can be written in three different forms: fraction, decimal, and
X-treme   with the percent symbol (%).

Example:   $0.50 $\frac{50}{100}$ 0.5   50% $\frac{1}{2}$ dollar

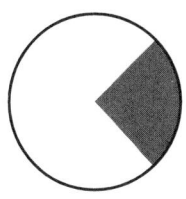 $\frac{1}{4}$ circle   0.25   25%

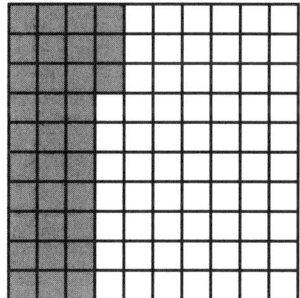 $\frac{33}{100}$ 0.33   33%

PRACTICE

Directions: Identify the percent given for the shaded part of each of the following.

1   _____ (%) percent

2   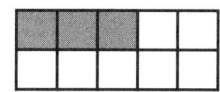   _____ (%) percent

3   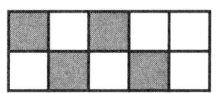   _____ (%) percent

4      _____ (%) percent

Directions: Shade the portion of the figure with the percent given.

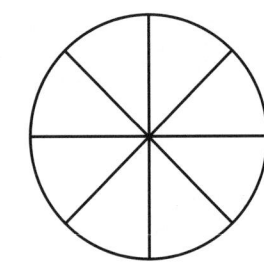

5   25%

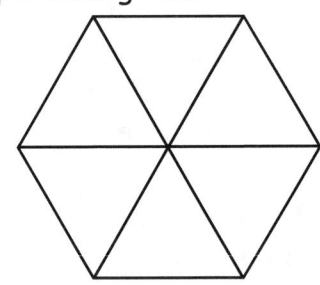

6   50%

2.6 CHANGING PERCENTS TO FRACTIONS AND DECIMALS

Remember: To turn a percent into a fraction, put it over 100 and reduce. To turn a percent into a decimal, move the decimal place (point) over to the left two places.

Example: Turn the following percents into fractions and decimals.

Percent	Fraction	Decimal
50 %	$\frac{50}{100} = \frac{1}{2}$	0.50
25 %	$\frac{25}{100} = \frac{1}{4}$	0.25
10 %	$\frac{10}{100} = \frac{1}{10}$	0.10
47 %	$\frac{47}{100}$	0.47

Hint: To change a fraction to a decimal, divide the numerator (the top number) by the denominator (the bottom number).

PRACTICE

Directions: Turn the following percents into fractions and decimals.

Percent	Fraction	Decimal
27 %		
16 %		
4 %		
68 %		
21 %		
87 %		

2.7 CALCULATING PERCENTS

Proportions can be used to solve percent problems. One proportion that is commonly used is:

$$\frac{part}{whole} = \frac{percent}{100}$$

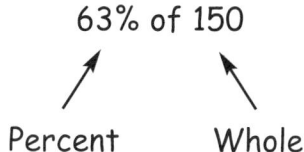 **Remember**: When working with percent problems, there are three situations:
1 When the part is missing
2 When the whole or base is missing
3 When the percent is missing

FINDING A NUMBER
Example: What is 63% of 150?

1) Identify what is missing.

$$63\% \text{ of } 150$$

Percent   Whole

2) Fill in the proportion; put x in place of what is missing.

$$\frac{x}{150} = \frac{63}{100}$$

3) Cross Multiply.

$$\frac{x}{150} = \frac{63}{100}$$

$$\frac{100x}{100} = \frac{9450}{100}$$

$$x = 94.5$$

PRACTICE

Directions: Find the missing number.

1   44% of 46 is what number?   _____

2   30% of 70 is what number?   _____

3   73% of 79 is what number?   _____

4   24% of 100 is what number?   _____

5   11% of 18 is what number?   _____

6   95% of 49 is what number?   _____

FINDING THE BASE

Example: 21 is 25% of what number?

1) Identify what is missing.

$$25\% \text{ of a number is } 21$$

Percent   Part

2) Fill in the proportion; put x in place of what is missing.

$$\frac{21}{x} = \frac{25}{100}$$

3) Cross Multiply.

$$\frac{21}{x} = \frac{25}{100}$$

$$\frac{2100}{25} = \frac{25x}{25} \quad \text{or} \quad \frac{25x}{25} = \frac{2100}{25}$$

$$x = 84$$

PRACTICE

Directions: Find the missing base.

1   36% of what number is 63?   _____

2   16% of what number is 5?   _____

3   88% of what number is 132?   _____

4   62% of what number is 775?   _____

5   4% of what number is 18?   _____

6   24% of what number is 24?   _____

FINDING THE PERCENT

Example: 15 is what percent of 75?

1) Identify what is missing.

15 is what percent of 75?

Part   Whole

2) Fill in the proportion; put x in place of what is missing.

$$\frac{15}{75} = \frac{x}{100}$$

3) Cross Multiply.

$$\frac{15}{75} = \frac{x}{100}$$

$$\frac{1500}{75} = \frac{75x}{75} \quad \text{or} \quad \frac{75x}{75} = \frac{1500}{75}$$

$$x = 20\ \%$$

PRACTICE

Directions: Find the percent.

1   20 is what percent of 25?   _____

2   What percent of 88 is 66?   _____

3   What percent is 15 of 30?   _____

4   What percent is 4 of 80?   _____

5   15 is what percent of 120?   _____

6   25 is what percent of 60?   _____

TEST PREP

1 Which ratio is equal to $\frac{45}{72}$?

    A $\frac{8}{3}$

    B $\frac{5}{8}$

    C $\frac{16}{6}$

    D $\frac{3}{7}$

2 Which is a rate?

    F $\frac{\$1.00}{\$10.00}$

    G $\frac{12}{24}$

    H $\frac{12 \text{ tokens}}{\$10.00}$

    J $\frac{2 \text{ tokens}}{14 \text{ tokens}}$

3 What percent of the figure is shaded?

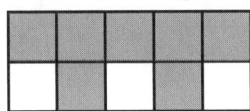

    A 30 %
    B 7%
    C 3 %
    D 70%

4 Which of the following shows a true proportion?

    F $\frac{4}{3} = \frac{12}{16}$

    G $\frac{5}{7} = \frac{10}{15}$

    H $\frac{22}{6} = \frac{11}{2}$

    J $\frac{18}{12} = \frac{3}{2}$

5 Find 65% of 80
    A 5200
    B 5.2
    C 52
    D 520

6 Two quarts of water weighs about 4 pounds. About how much does 4 quarts of water weigh?
    F 1 lb
    G 4 lbs
    H 6 lbs
    J 8 lbs

7 Write 0.386 as a percent.
    A 38.6%
    B 386%
    C 3.86%
    D 0.00386%

8 Solve: $\frac{6}{8} = \frac{x}{160}$

    F 20
    G 60
    H 120
    J 960

9 Write $\frac{12}{25}$ as a percent.

    A 48 %
    B 48
    C 0.48%
    D 12%

10 Write 72% as a fraction in simplest form.

    F $\frac{72}{100}$

    G $\frac{18}{25}$

    H $\frac{36}{50}$

    J 18%

11 Which does not show the ratio 2 apples to 3 oranges?

A $\dfrac{2}{3}$

B 2 to 3

C 2:3

D $\dfrac{3}{2}$

12 Which figures appear to be similar?

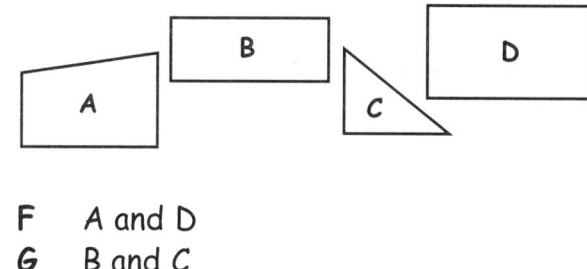

F A and D
G B and C
H A and C
J B and D

13 Adam sells computer hardware. He earns 27% of his total sales in commission. If he earned $900 in commissions last month, what were his total sales for the month? **Show your work.**

Answer:_____

14 The middle school math classes were collecting cans for the local food drive. Their goal was 1,000 cans. So far, 550 cans have been collected. What percent is this of the goal? **Show your work.**

Answer:_____

15 Find the missing sides of the triangles below. ΔXYZ ~ ΔEFG. [not drawn to scale] **Show your work.**

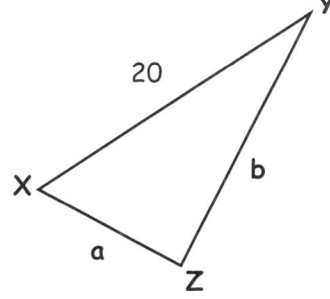

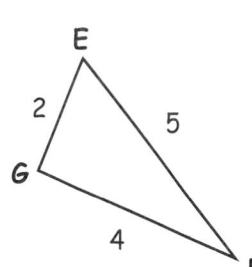

a _____

b _____

16 Solve the following proportion by cross multiplying.

Show your work.

$$\frac{20}{8} = \frac{n}{2}$$

Answer:_____

17 Molly bought a DVD that was priced at $19.99. She had to pay 8.25% sales tax. How much sales tax did she have to pay?

Show your work.

Answer:_____

18 The lengths of the sides of a triangle are 45 inches, 55 inches, and 70 inches. The longest side of a similar triangle is 42 inches. What are the lengths of the other two sides of the similar triangle? [not drawn to scale]

Show your work.

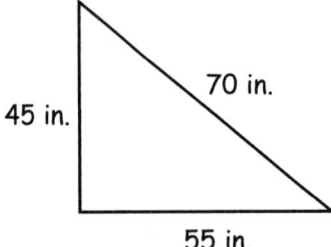

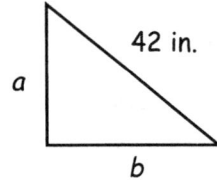

Answer *a*: _____

Answer *b*: _____

Vocabulary
These words and phrases are associated with Operations and may be used when answering questions in this chapter. Definitions can be found in the Glossary/Index at the back of this *X-treme Review*.

base	fraction	reciprocal
equivalent fraction	improper fraction	order of operations
exponent	least common denominator	power
exponential form	mixed number	unlike denominators

A New Art

Started in the first century – about 100 A.D., origami actually originated in China. It was not until the sixth century that it moved to Japan where it became a popular art form. The Japanese further developed origami to the point that it is more commonly recognized as a Japanese art. Simply put, origami is the art of folding paper to form three-dimensional figures. Both the knowledge of fractions and geometry are used when making origami figures. The final figures often include the multiple folding of a piece of paper in quarters , thirds, halves, or any number of common fractions.

Try making a paper cup by following these five steps:

(1)

1 Fold a square sheet in half along the dotted line.

(2)

2 Fold the corner up to halfway along the opposite side as shown.

(3)
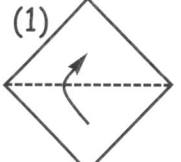

3 Fold the other corner up in the same way.

(4)
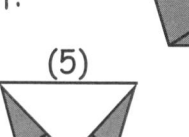

4 Split the top flaps and fold one down on each side to get:

(5)

5 Your finished cup.

In this lesson, you will review how to perform four operations with fractions. You will also review how to perform those operations in a specific order.

http://www.netguides.org.uk/guides/origami/introduction.html
http://library.thinkquest.org/5402/history.html

LESSON 3: OPERATIONS

3.1 FRACTIONS AND MIXED NUMBERS

A (1) _____ is a number that represents part of a whole, part of a set, or a quotient in the form $\frac{a}{b}$ which can be read as *a divided by b*.

(2) _____ are two or more fractions that have the same quotient or that name the same region, part of a set, or part of a segment.

Example 1: $\frac{1}{3} = \frac{1 \times 3}{3 \times 3} = \frac{3}{9}$ $\frac{1}{2} = \frac{1 \times 4}{2 \times 4} = \frac{4}{8}$

A (3) _____ is a number composed of a whole number and a proper fraction.

An (4) _____ is a fraction whose numerator is larger than its denominator.

Remember: To change a mixed number to an improper fraction, follow these steps:
X-treme
1 Multiply the denominator by the whole number.
2 Add the numerator.
3 The sum is the numerator of the improper fraction and the denominator is the same.

Example 2: $3\frac{2}{5}$ ⟶ $5 \times 3 = 15$

$3\frac{2}{5}$ ⟶ $15 + 2 = 17$

So: $3\frac{2}{5} = \frac{17}{5}$

Remember: To change an improper fraction to a mixed number, divide the numerator by
X-treme   the denominator. The quotient is equal to the whole number and the remainder is the new numerator over the same denominator.

Example 3: $\frac{7}{3}$

$3\overline{)7}$ ⟶ whole number (plus a remainder)

So: $\frac{7}{3} = 2\frac{1}{3}$

PRACTICE

Directions: Write two fractions equivalent to each fraction given.

1   $\frac{3}{5}$ _____

2   $\frac{8}{12}$ _____

3   $\frac{2}{7}$ _____

Directions: Write each mixed number as an improper fraction.

4   $1\frac{1}{6}$ _____

5   $4\frac{7}{10}$ _____

6   $8\frac{2}{3}$ _____

Directions: Write each improper fraction as a mixed number.

7   $\frac{11}{6}$ _____

8   $\frac{23}{10}$ _____

9   $\frac{27}{5}$ _____

3.2 ADD AND SUBTRACT FRACTIONS AND MIXED NUMBERS

Two fractions with different denominators have (1) _____.

Remember: To add or subtract fractions with unlike denominators, change them to equivalent fractions with the same denominator. You can find equivalent fractions by either multiplying or dividing the numerator and the denominator of a fraction by the same number. This is known as the (2) _____ or LCD, which is also the least common multiple of the two denominators.

Example 1:

$\frac{1}{4} + \frac{1}{2}$

$\frac{1}{4} = \frac{1}{4}$ ⟶ It already has a denominator of 4.

$\frac{1}{2} = \frac{2}{4}$ ⟶ Multiply numerator and denominator by 2 to make the denominator 4.

$+$ ___ ⟶ Add the two resulting fractions.

$\frac{3}{4}$ ⟶ Answer: $\frac{1}{4} + \frac{1}{2} = \frac{3}{4}$

Example 2: $\dfrac{7}{15} - \dfrac{3}{10}$

The least common multiple of 15 and 10 is 30. So, 30 is the LCD.

$\dfrac{7}{15} = \dfrac{14}{30}$ $\longrightarrow$ Multiply numerator and denominator by 2.

$\dfrac{3}{10} = \dfrac{9}{30}$ $\longrightarrow$ Multiply numerator and denominator by 3.
$\longrightarrow$ Subtract the two resulting fractions.

$\dfrac{5}{30} = \dfrac{1}{6}$ $\longrightarrow$ Write the answer in lowest terms.

So: $\dfrac{7}{15} - \dfrac{3}{10} = \dfrac{1}{6}$

Remember: To add mixed numbers, find the fractions that have a common denominator, add the fractions, then add the whole numbers.

Example 3: $2\dfrac{3}{10} + 1\dfrac{1}{4}$

$2\dfrac{3}{10} = 2\dfrac{6}{20}$ $\longrightarrow$ Multiply numerator and denominator by 2.

$1\dfrac{1}{4} = 1\dfrac{5}{20}$ $\longrightarrow$ Multiply numerator and denominator by 5.
$\longrightarrow$ Add the two resulting fractions.

$3\dfrac{11}{20}$

So: $2\dfrac{3}{10} + 1\dfrac{1}{4} = 3\dfrac{11}{20}$

Example 4: $5\dfrac{1}{2} = 5\dfrac{2}{4}$

$2\dfrac{3}{4} = 2\dfrac{3}{4}$ $\longrightarrow$ Add the two resulting fractions.

$7\dfrac{5}{4}$

$7\dfrac{5}{4}$ $\longrightarrow$ Simplify: $7\dfrac{5}{4} = 7 + 1\dfrac{1}{4} = 8\dfrac{1}{4}$

So: $7\dfrac{5}{4} = 8\dfrac{1}{4}$

Remember: To subtract mixed numbers, write equivalent fractions with a common denominator, subtract the fractions, then subtract the whole numbers.

Example 5: $3\frac{2}{3} - 2\frac{1}{4}$

$3\frac{2}{3} = 3\frac{8}{12}$ ⟶ Multiply numerator and denominator by 4.

$2\frac{1}{4} = 2\frac{5}{12}$ ⟶ Multiply numerator and denominator by 3.
⟶ Subtract the two resulting fractions.

$1\frac{3}{12}$

Remember: Sometimes you will need to <u>rename</u> before subtracting. (Renaming involves multiplying or dividing both the numerator and the denominator of a fraction by the same number, giving <u>another</u> <u>name</u> for the same fraction.)

Example 6: $4\frac{1}{6} - 1\frac{5}{6}$

$4\frac{1}{6} = 3\frac{7}{6}$ ⟶ Since $\frac{1}{6} < \frac{5}{6}$, then $4\frac{1}{6}$ must be renamed.

Renaming: $4\frac{1}{6} = 3 + 1\frac{1}{6}$

$\quad\quad\quad\quad = 3 + \frac{7}{6} = 3\frac{7}{6}$

$1\frac{5}{6} = 1\frac{5}{6}$

⟶ Subtract the two resulting fractions.

$2\frac{2}{6} = 2\frac{1}{3}$ ⟶ Simplify.

Example 7: $8 - 3\frac{5}{8}$

$8 = 7\frac{8}{8}$ ⟶ Since $8 = 7 + 1$

$\quad\quad\quad = 7 + \frac{8}{8} = 7\frac{8}{8}$

$3\frac{5}{8} = 3\frac{5}{8}$ ⟶ Multiply numerator and denominator by 3.
⟶ Subtract the two resulting fractions.

$4\frac{3}{8}$

PRACTICE

Directions: Add or subtract. Write the answer in simplest form.

1 $\frac{1}{2} + \frac{1}{6}$ = _____

2 $\frac{1}{9} + \frac{2}{3}$ = _____

3 $4\frac{2}{3} + 1\frac{1}{4}$ = _____

4 $5\frac{4}{5} + 2\frac{2}{3}$ = _____

5 $\frac{1}{2} - \frac{1}{3}$ = _____

6 $\frac{4}{5} - \frac{2}{3}$ = _____

7 $6\frac{3}{8} - 3\frac{1}{4}$ = _____

8 $15\frac{1}{2} + 7\frac{2}{3}$ = _____

3.3 MULTIPLY AND DIVIDE FRACTIONS AND MIXED NUMBERS

Remember: To multiply fractions, multiply the numerators and multiply the denominators. Rewrite the answer in lowest terms.

Example 1: $\frac{2}{3} \times \frac{1}{5} = \frac{2}{15}$

Remember: There is a shortcut for multiplying fractions. Sometimes you can simplify the problem before you multiply.

Example 2: $\frac{1}{\overset{2}{\underset{1}{2}}} \times \frac{\overset{2}{4}}{5} = \frac{2}{5}$

Remember: To multiply mixed numbers, first write each mixed number as an improper fraction. Then, multiply the two improper fractions.

Example 3: $1\frac{1}{3} \times 2\frac{3}{4} = \frac{4}{3} \times \frac{11}{4}$

$\frac{\overset{1}{4}}{3} \times \frac{11}{\underset{1}{4}} = \frac{11}{3} = 3\frac{2}{3}$

Example 4: $4 \times 1\frac{3}{5} = \frac{4}{1} \times \frac{8}{5}$

$\frac{4}{1} \times \frac{8}{5} = \frac{32}{5} = 6\frac{2}{5}$

Two numbers are (1) _____ if their product is 1.

Remember: To find the reciprocal of a fraction, exchange or "flip" the numerator and *Xtreme* denominator.

Example 5: The reciprocal of $\frac{3}{4}$ is $\frac{4}{3}$.

Remember: To divide fractions, multiply by the reciprocal (reciprocal of $\frac{2}{5}$ is $\frac{5}{2}$) of the *Xtreme* divisor (the second fraction).

Example 6: $\frac{3}{8} \div \frac{2}{5}$

$$\frac{3}{8} \times \frac{5}{2} = \frac{15}{16}$$

reciprocal

Remember: To divide mixed numbers, first write each mixed number as an improper *Xtreme* fraction, then multiply by the reciprocal of the divisor.

Example 7: $3\frac{3}{4} \div 1\frac{4}{5} = \frac{15}{4} \div \frac{9}{5}$

Example 8: $\frac{\overset{5}{\cancel{15}}}{4} \times \frac{5}{\underset{3}{\cancel{9}}} = \frac{25}{12} = 2\frac{1}{12}$

PRACTICE

Directions: Multiply or divide. Write the answer in lowest terms.

1 $\frac{6}{7} \times \frac{1}{5}$ = _____ 2 $\frac{5}{8} \times \frac{4}{5}$ = _____ 3 $\frac{7}{10} \div \frac{1}{2}$ = _____

4 $1 \div \frac{2}{9}$ = _____ 5 $8\frac{1}{2} \times 2\frac{1}{3}$ = _____ 6 $5\frac{1}{5} \times 4\frac{5}{6}$ = _____

7 $2\frac{2}{3} \div 1\frac{5}{9}$ = _____ 8 $6\frac{2}{5} \div 2\frac{4}{5}$ = _____

3.4 Exponents

An (1) _____ is used to show repeated multiplication. The exponent specifies how many times the (2) _____ is being used as a factor.

In 3^4, the 4 is the exponent. It tells that 3 is to be used as a factor 4 times.

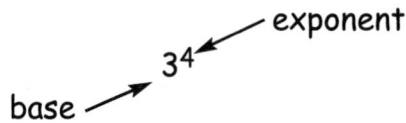

exponent

3^4

base

Remember: 3^4 is read, "3 to the fourth power."

exponential form

simplified

$$3^4 = \underline{3 \times 3 \times 3 \times 3} = 81$$

repeated multiplication

Practice

Directions: Write each of the following as repeated multiplication.

1 4^2 _____

2 7^4 _____

3 9^3 _____

4 2^4 _____

Directions: Use exponents to rewrite each of the following.

5 $8 \times 8 \times 8$ _____

6 $5 \times 5 \times 5 \times 5$ _____

7 10×10 _____

8 $6 \times 6 \times 6 \times 6 \times 6$ _____

Directions: Simplify each of the following.

9 $3^2 =$ _____

10 $5^3 =$ _____

11 $4^1 =$ _____

12 $2^3 =$ _____

3.5 ORDER OF OPERATIONS

A set of standard rules known as (1) _____ is used so that everyone gets the same answer when calculating an expression.

ORDER OF OPERATIONS
1   Do any work inside the parentheses.
2   Calculate exponents.
3   Multiply and divide as they appear from left to right.
4   Add and subtract as they appear from left to right.

Remember: When doing order of operations, work your way down step-by-step.

X-treme

Example 1:

$14 + \underline{5 \cdot 2}$ ⟶ Do any multiplication first.

$14 + 10$ ⟶ Then, recopy any part of the expression on which you have not worked.

$\underline{14 + 10} = 24$ ⟶ Finally, add.

Example 2:

$100 \div \underline{(25 - 5)} + 4$ ⟶ Do any parentheses first.

$100 \div 20 + 4$ ⟶ Recopy.

$\underline{100 \div 20} + 4$ ⟶ Do the division.

$\underline{5 + 4} = 9$ ⟶ Add.

Example 3: $48 - 3^2 \cdot \underline{(8 \div 2)} + 10$ ⟶ Do any parentheses first.

$48 - \underline{3^2} \cdot 4 + 10$ ⟶ Calculate the exponent.

$48 - \underline{9 \cdot 4} + 10$ ⟶ Multiply.

$\underline{48 - 36} + 10$ ⟶ Subtract (because it appears first).

$\underline{12 + 10} = 22$ ⟶ Add.

PRACTICE

Directions: Simplify each expression.

1 $9 + 24 \div 3 =$

2 $5^2 + (16 \div 8)$

3 $18 \div 6 \cdot 2 - (24 \div 2^3)$

4 $(10 - 3 \cdot 2) + 4^2 \div 8$

TEST PREP

1 Which is equivalent to $\frac{4}{5}$?

    A $\frac{6}{7}$

    B $\frac{12}{15}$

    C $\frac{4}{10}$

    D $\frac{2}{10}$

2 Which is in lowest terms?

    F $\frac{5}{8}$

    G $\frac{6}{15}$

    H $\frac{9}{18}$

    J $\frac{8}{14}$

3 What is the least common denominator for $\frac{5}{6}$ and $\frac{3}{8}$?

    A 48

    B 24

    C 12

    D 16

4 Which of the following shows the improper fraction for $6\frac{4}{7}$?

    F $\frac{10}{7}$

    G $\frac{42}{7}$

    H $\frac{46}{7}$

    J $\frac{64}{7}$

5 Which shows an equivalent fraction for $\frac{5}{8}$ and an equivalent fraction for $\frac{5}{6}$, using the LCD?

A $\frac{15}{24}; \frac{20}{24}$

B $\frac{30}{48}; \frac{40}{48}$

C $\frac{5}{24}; \frac{20}{24}$

D $\frac{10}{24}; \frac{20}{24}$

6 Write $\frac{31}{8}$ as a mixed number.

F 4

G $3\frac{7}{8}$

H $\frac{7}{8}$

J $4\frac{3}{8}$

7 Find $\frac{5}{6} + \frac{1}{4}$ in simplest form.

A $\frac{2}{5}$

B $\frac{11}{12}$

C $1\frac{1}{4}$

D $1\frac{1}{12}$

8 Adrian writes for $\frac{3}{10}$ of an hour in the morning and $\frac{1}{5}$ of an hour in the afternoon. How long does he write in all?

F $\frac{1}{10}$ hour

G $\frac{1}{2}$ hour

H $\frac{1}{5}$ hour

J $\frac{2}{5}$ hour

9 Kimli used $2\frac{4}{9}$ feet of ribbon to trim a pillowcase and $6\frac{2}{3}$ feet of ribbon to trim a quilt. How much ribbon did she use in all?

A $8\frac{1}{2}$

B $8\frac{3}{5}$

C $9\frac{1}{9}$

D $9\frac{1}{10}$

10 Find $\frac{7}{9} - \frac{2}{3}$ in lowest terms.

F $\frac{1}{9}$

G $\frac{1}{6}$

H $\frac{5}{9}$

J $\frac{5}{6}$

11 Bill baked $6\frac{2}{3}$ dozen cookies for a bake sale, and $3\frac{1}{4}$ dozen of the cookies were sold. How many dozen cookies were left over?

A $3\frac{5}{12}$

B $3\frac{7}{12}$

C $4\frac{5}{12}$

D $4\frac{7}{12}$

12 A recipe for granola bars calls for $\frac{2}{3}$ cup of oats. How much of the oats would you use to make $\frac{1}{2}$ of the amount in the original recipe?

F $\frac{3}{5}$

G $\frac{2}{5}$

H $\frac{2}{6}$

J $\frac{3}{6}$

13 Antonio has $2\frac{2}{3}$ of a piece of wood that he wants to cut into 4 pieces. How long should he cut each piece?

A $8\frac{2}{3}$

B $10\frac{2}{3}$

C $\frac{8}{3}$

D $\frac{2}{3}$

14 Simplify: $7 + 24 \div (-2)$

F 19

G -5

H 15.5

J 5

15 Evaluate $(1 + 4 \cdot 2)^2$

A 18

B 100

C 66

D 81

16 What is the square of 14?

F 226

G 196

H 169

J 28

17 Which expression is equal to 2?

A $24 \div (2 + 2 \cdot 5)$

B $(24 \div 2 + 2) \cdot 5$

C $(24 \div 2) + (2 \cdot 5)$

D $24 \div (2 \cdot 2) \cdot 5$

18 Write 100 as a power of 10

   F 10^1

   G 100^1

   H 10^2

   J 100^2

19 Which expression represents $5 \times 5 \times 5 \times 5$ in exponential form?

   A 4^5

   B 5^4

   C   625

   D   20

20 First-class postage in the United States costs 39¢ for 1 ounce. Your package weighs $\frac{3}{4}$ oz. Do you need extra postage to include a letter that weighs $\frac{3}{8}$ oz.? Explain.

Answer: _____

_____

_____

_____

_____

21 You are hiking a trail that is $13\frac{1}{2}$ miles long. You plan to rest every $2\frac{1}{4}$ miles. How many stops will you make?

Show your work.

Answer: _____

22 Simplify the following expression.

Show your work.

$$2 \cdot 5^2 + 4 \cdot 2^3 - 6(5 - 3)$$

Answer: _____

LESSON FOUR

ALGEBRA

Vocabulary

These words and phrases are associated with Algebra and may be used when answering questions in this chapter. Definitions can be found in the Glossary/Index at the back of this *X-treme Review*.

equation	expression	variable
evaluate	formula	

Learning to Speak a New Language

Have you ever wished that someday you would be able to speak two or even three languages? Well, from day one in Kindergarten, you have been learning a new language called Math. Until now, you have been working with the background of the language, the numbers. Now, you are just beginning to work with a very important part of the language, Algebra. Still not sure that Algebra is a language? *Check this out!*

The ABC Basics of Language

Any language needs an alphabet. In Algebra, we use the English alphabet for our variables. In any language, you write expressions and sentences. In Algebra, you write numerical expressions and sentences with or without variables.

Have you ever wondered why your teacher has been using the term, "translating"? You have been learning to change between languages! So, now you can tell everyone that you can communitcate in at least two languages!

In this lesson you will review how to translate algebraic expressions and how to evaluate them. You will also solve basic one-step equations using basic whole number facts.

LESSON 4: ALGEBRA

4.1 VERBAL EXPRESSIONS INTO ALGEBRAIC EXPRESSIONS

An (1) _____ is a mathematical representation containing numbers, variables, and operating symbols.

Example: $(5 + 2) - 27 \div 3$ ⟶ numerical expression

   $2x + 3y$ ⟶ algebraic expression

A (2) _____ is a symbol used to represent a number or group of numbers in an expression or an equation.

Remember: You can often use symbols to translate word phrases into mathematical Xtreme expressions. The following chart contains some of the word phrases that you can associate with each of the four operations.

+	–	×	÷
add	subtract	multiply	divide
sum	difference	product	quotient
plus	minus	times	for
total	remainder	of	per
more than	less than		
increased by	decreased by		
added to	take away		
and	fewer		

Remember: When you see the word "than" in a word phrase, <u>reverse</u> its meaning. Use Xtreme add (+) for "less than" and use subtract (–) for "greater than." For example, "4 less <u>than</u> y" means "y – 4" or "4 greater <u>than</u> y" means "y + 4.".

Example: Write a variable expression for each of the following word phrases:

1 A number x increased by ten.
   Expression: $x + 10$
   } The phrase, *increased by*, means addition.

2 Thirteen less than a number y.
   Expression: $y - 13$
   } The phrase, *less than*, means subtraction.

3 A number b decreased by seven, divided by twenty.
   Expression: $(b - 7) \div 20$
   } The phrases, *decreased by* and *divided by*, mean two operations – subtraction and division.

4 The product of fifteen and the sum of eight and a number m.
   Expression: $15(8 + m)$
   } The words, *product* and *sum*, mean two operations – multiplication and addition.

PRACTICE

Directions: Write a variable expression for each word phrase.

1   The product of nine and a number c. _____

2   A number f subtracted from twelve. _____

3   The quotient when a number n is divided by 11
    decreased by 3. _____

4   The sum of a number x and four, multiplied
    by 25. _____

5   The product of seventeen less than a number
    y and the sum of a number z and 21. _____

4.2 EVALUATING ALGEBRAIC EXPRESSIONS

To (1) _____ means to find the value of a mathematical expression.

Remember: You can evaluate an algebraic expression by replacing or substituting, each
X-treme   variable with a number. Don't forget to use Order of Operations to simplify.

Example 1: Evaluate $3y - 12$, where $y = 9$.

$$3y - 12$$

$$3(9) - 12 \longrightarrow \text{Replace } y \text{ with 9.}$$

$$27 - 12$$

$$15$$

Example 2: Evaluate $6(a + 4)$, where $a = 2$.

$$6(a + 4)$$

$$6(2 + 4) \longrightarrow \text{Replace } a \text{ with 2.}$$

$$6(6)$$

$$36$$

Example 3: Evaluate $4c - d$, where $c = 3$ and $d = 4$.

$$4c - d$$

$$4(3) - 4 \longrightarrow \text{Replace } c \text{ with 3 and } d \text{ with 4.}$$

$$12 - 4$$

$$8$$

PRACTICE

Directions: Evaluate each expression.

1 $2a + 5$, where $a = 5$.

2 $19 - (b - 4)$, where $b = 8$.

3 $13xy$, where $x = 1$ and $y = 7$.

4 $16 - 4mn$, where $m = 0$ and $n = 2$.

5 abc, where $a = 5$, $b = 6$, and $c = 7$.

6 $4x - y + \dfrac{x}{2}$, where $x = 3$ and $y = 4$

4.3 EVALUATING FORMULAS

A (1) _____ is a mathematical statement, equation, or rule that shows a relationship between two or more quantities.

Remember: Evaluating a given formula is the same as evaluating an algebraic expression. Substitute given values for the appropriate variables.

X-treme

Example 1: Use the distance formula, $d = rt$, to solve for d, $r = 45 \text{ mi.}/_{\text{hr.}}$ and $t = 3$ hr., where d = distance, r = rate, and t = time.

$$d = rt \longrightarrow \text{Write the formula.}$$

$$d = (45 \text{ mi.}/_{\text{hr.}})(3 \text{ hr}) \longrightarrow \text{Substitute given values.}$$

$$d = 135 \text{ mi.} \longrightarrow \text{Simplify.}$$

Example 2: Given the formula for the volume of a box: $V = l \cdot w \cdot h$ (length, width, height), find the volume, if $l = 2$cm, $w = 3$ cm, and $h = 4$ cm.

$V = l \cdot w \cdot h$ ⟶   Write the formula.

$V = 2 \cdot 3 \cdot 4$ ⟶   Substitute given values.

$V = 24$ cm^3 ⟶   Simplify.

PRACTICE

Directions: Use the distance formula, $d = rt$. Find the missing value.

1 $r = 24$ mi./hr., $t = 1$ hr., $d =$ _____

2 $r = 50$ mi./hr., $t = 30$ min., $d =$ _____

3 $r = 54$ mi./hr., $t = 6$ hr., $d =$ _____

4 $r = 12$ cm/hr., $t = 0.5$ hr., $d =$ _____

Directions: Use the formula: $V = lwh$ to find the surface area for a triangular prism.

5 $l = 9$ cm
    $w = 6$ cm
    $h = 4$ cm

    Volume = _____ cm^3

6 $l = 8$ in.
    $w = 3$ in.
    $h = 5$ in.

    Volume = _____ in.^{3}

4.4 SOLVING ONE-STEP EQUATIONS

An (1) _____ is a mathematical sentence stating that two expressions are equal.

Remember: To solve an equation, you "undo" operations until the variable is alone on one side of the equal sign.

Example 1: $x + 7 = 18$

$$
\begin{array}{rl}
x + 7 &= 18 \\
-7 &\ -7 \\
\hline
x &= 11
\end{array}
$$
⟶ Undo addition by subtracting 7 on both sides.

Example 2: $y - 3 = 10$

$$\begin{array}{r} y - 3 = 10 \\ \underline{+\ 3 \quad +\ 3} \\ y \quad\ \ = 13 \end{array}$$ $\longrightarrow$ Undo subtraction by adding 3 to both sides.

Example 3: $5a = 20$

$$\frac{5a}{5} = \frac{20}{5}$$ $\longrightarrow$ Undo multiplication by dividing each side by 5.

$$a = 4$$

Example 4: $\dfrac{b}{6} = 8$

$$6 \cdot \frac{b}{6} = 8 \cdot 6$$ $\longrightarrow$ Undo division by multiplying each side by 6.

$$b = 48$$

PRACTICE

Directions: Solve each equation. **Show your work!**

1 $x + 6 = 14$ **2** $m - 10 = 7$ **3** $y + 11 = 23$

4 $a - 12 = 2$ **5** $9b = 72$ **6** $3n = 108$

7 $\dfrac{c}{5} = 25$ **8** $\dfrac{x}{4} = 13$

TEST PREP

1 Which expression means x less than 4?

   A $x + 4$

   B $x - 4$

   C $4 - x$

   D $x \div 4$

2 Evaluate $12 + x$, where $x = 4$

   F   16

   G   8

   H   3

   J   48

3 Solve: $x - 8 = 28$

   A   6

   B   8

   C   36

   D   20

4 Solve: $\frac{m}{4} = 7$

   F   68

   G   12

   H   3

   J   28

5 Write an expression for the following phrase: "The sum of n and twenty-two, multiplied by three."

   A $3n + 22$

   B $n + 3 \cdot 22$

   C $3 - n + 22$

   D $3(n + 22)$

6 Rose sells earrings for $30 a pair. Which expression shows how much Rose will earn if she sells e pairs of earrings?

   F $30 + e$

   G $30e$

   H $e \div 30$

   J $30(e + 1)$

7 Use the distance formula $d = rt$. How many miles would a car go, traveling 65 miles per hour in 3.2 hours?

   A   208 miles

   B   195 miles

   C   416 miles

   D   104 miles

8 When $x = 13$ and $y = 7$, which expression is equal to 39?

   F $2(x + y) - 1$

   G $x + 2y$

   H $2x + y$

   J $3x - 3y$

9 What is the volume of the box shown below?

$$V = l \cdot w \cdot h$$

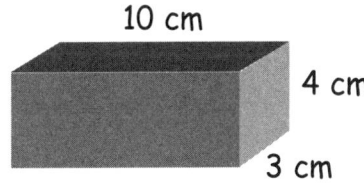

   A $17 \ cm^3$

   B $120 \ cm^3$

   C $52 \ cm^3$

   D $70 \ cm^3$

10 Solve: $12x = 156$

    A 144

    B 1870

    C 168

    D 13

11 At the local soft drink factory, 50 bottles are filled per minute. How many bottles are filled in $1\frac{1}{2}$ hours? **Show your work.**

Answer: _____ bottles

12 Tell whether the given value is a solution to the equation. Explain your answer.

$$x - 3 = 21, \; x = 18$$

Answer: _____

_____

_____

_____

_____

13 A certain number decreased by 9 is equal to the product of 4 and 5. Find the number. **Show your work.**

Answer: _____

14 The formula for converting Celsius temperature to Fahrenheit is: $F° = \dfrac{9}{5} C + 32$. The thermometer reads 25° C. What is the temperature in degrees Fahrenheit?
Show your work.

Answer: _____ ° F

LESSON FIVE
COORDINATE GEOMETRY

5
X-treme

Vocabulary

These words and phrases are associated with Coordinate Geometry and may be used when answering questions in this chapter. Definitions can be found in the Glossary/Index at the back of this *X-treme Review*.

coordinate geometry	plot	quadrilateral	triangle
coordinate plane	point	rectangle	vertex and vertices
ordered pair	parallelogram	rhombus	x-axis
origin	geometric shape	square	x-y coordinate plane
perimeter	quadrant	trapezoid	y-axis

Plotting the Course of a Hurricane.

Hurricane Katrina (satellite photo) was one of the most costly, severe storms to hit the mainland of the United States, especially the Gulf Coast from Pensicola, FL to New Orleans, LA. Fortunately, using satellites, radar, and even reconnaissance aircraft, the National Hurricane Center can plot the course (track) of a hurricane. Also, scientists can predict when and where it will "hit" land, how strong it will be, and when it will get to that location.

This grid shows the track of Hurricane Katrina, using coordinate geometry.

http://www.nhc.noaa.gov/
http://www.liv-redcross.org/katrina-rita-statistics.htm

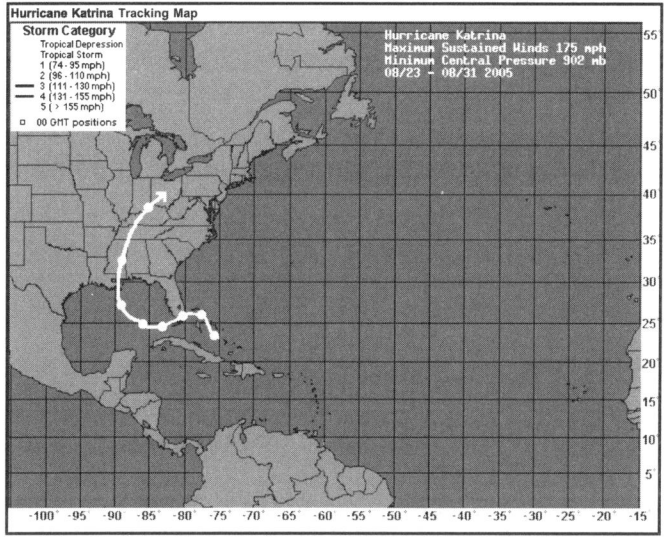

In this lesson you will review working in Quadrant I of the coordinate system. You will also review using the coordinate plane to create and classify polygons, as well as to calculate their perimeter and area.

American Red Cross Katrina Facts

- Red Cross relief exceeded $2 billion in contributions.
- 1.2 million families needed financial assistance.
- More than 3.7 million victims survived Hurricane Katrina.
- More than 219,500 Red Cross disaster relief workers from all 50 states responded to their neighbors in need.
- More than 27.4 million hurricane survivors received hot meals and 25.2 million snacks.
- More than 3.42 million stayed overnight in nearly 1,100 shelters across 27 states and the District of Columbia.

LESSON 5: COORDINATE GEOMETRY

5.1 COORDINATE GEOMETRY

(1) _____ is the study of geometry using a coordinate plane.

The (2) _____ is a plane containing a set of coordinate axes.

The (3) _____ is the horizontal number line. The

(4) _____ is the vertical number line.

The point at which both number lines intersect is called the (5) _____.

The intersection of the number lines divide the plane into four (6) _____.

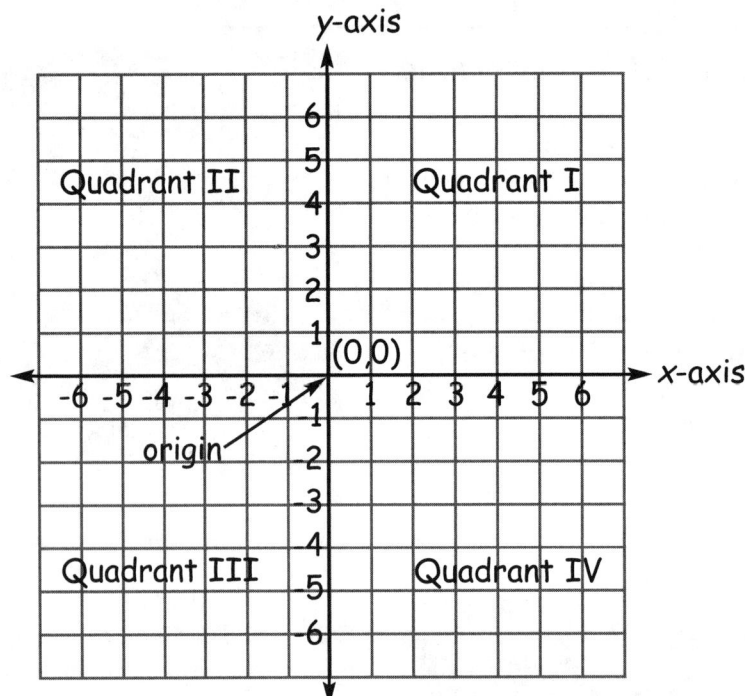

An (7) _____ gives the coordinates and location of a point.

Remember:

(*x*, *y*)

The *x*-coordinate tells how far to the left or right of the origin the point is located.

The *y*-coordinate tells how far up or down the point is located.

Example: Graph the point P(4,5)

1 Start at the origin.
2 Move 4 places to the right.
3 Move 5 places up the grid.

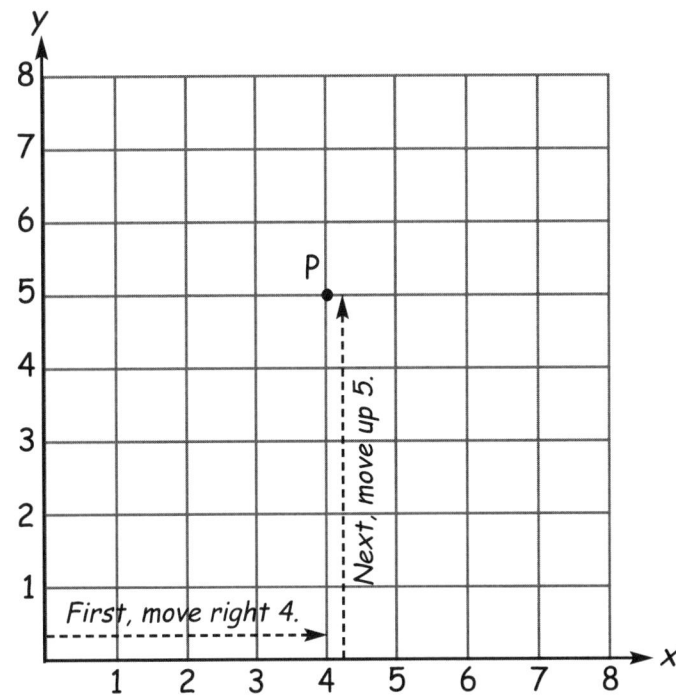

PRACTICE

Directions: Use the graph of Quadrant I below to give the coordinates of each point.

1 A – (____, ____)

2 B – (____, ____)

3 C – (____, ____)

4 D – (____, ____)

5 E – (____, ____)

6 F – (____, ____)

Directions: Name the point with the given coordinates using the same graph as used in 1 - 6..

7 (4, 2) _____

8 (15, 14) _____

9 (8, 8) _____

10 (5, 10) _____

11 (10, 2) _____

12 (2, 14) _____

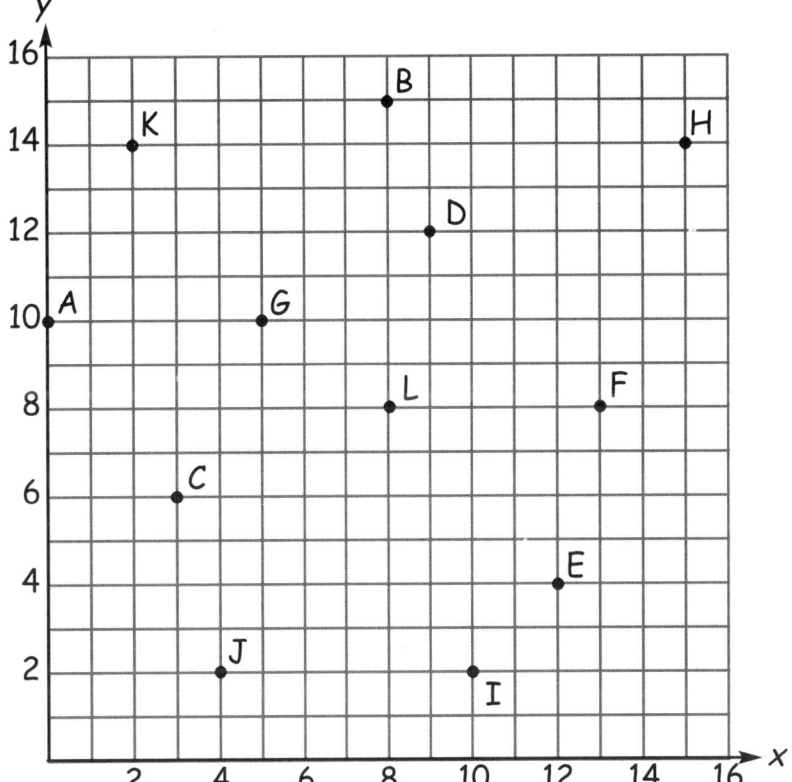

5.2 POLYGONS

A (1) _____ is any regular or irregular polygon, circle, or combination of geometric figures. Polygons can be classified by their sides and angles.

Name of Polygon	Picture	Number of Sides	Number of Angles
Triangle		3	3
Quadrilateral		4	4
Pentagon		5	5
Hexagon		6	6

Remember: Some quadrilaterals have specific classifications.
X-treme

PRACTICE

Directions: Using the definition and picture, name each quadrilateral.

Name of Quadrilateral	Definition	Picture
1 _____	It is a rectangle with two adjacent sides congruent (all four sides are congruent).	
2 _____	It is a quadrilateral with four right angles.	
3 _____	It is a quadrilateral with two pairs of parallel sides.	
4 _____	It is a parallelogram with two adjacent sides congruent (all four sides are congruent).	
5 _____	It is a quadrilateral with exactly one pair of parallel sides.	

Example: Classify the following Polygon in as many ways possible.

This polygon is a **quadrilateral**, because it has four sides. It is also called a **trapezoid**, because it has exactly one pair of parallel sides. It has 4 vertices (in this case, corners) where the sides meet.

5.3 PERIMETER

(1) _____ is the distance around the outside of a polygon.

Remember: When looking for the perimeter of a polygon that is on a coordinate plane, *Xtreme* count the number of units to find the lengths of each side.

Example: Graph the following points in Quadrant I of the coordinate plane: (1, 1), (4, 3), (4, 3), and (4, 1). Connect the points in order and connect the last point to the first point. Find the perimeter of the figure drawn.

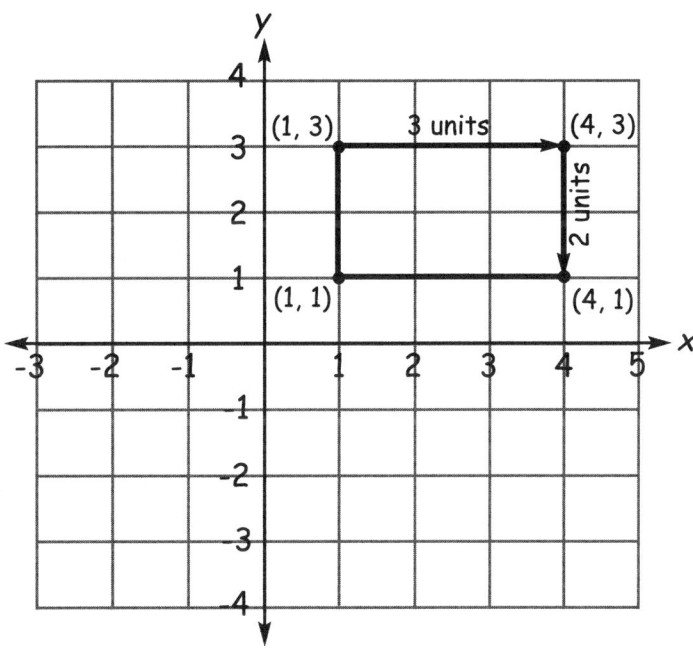

Problem: To find the perimeter (P) of the rectangle formed, you need to know the length (*l*) and width (*w*).

$$P = 2l + 2w$$

Solution: Since horizontal lines have the same *y*-coordinates, count the number of units between (1, 3) and (4, 3). There are 3 units in this case.

Since vertical lines have the same *x*-coordinates, count the number of units between (4, 3) and (4, 1). In this case, there are 2 units.

So,

P = 2*l* + 2*w*	⟶ Write the formula.
P = 2·3 + 2·2	⟶ Substitute the values you determined.
P = 6 + 4	⟶ Simplify.
P = 10 units	⟶ Perimeter.

PRACTICE

Directions: Graph the points and connect them with straight lines forming a polygon. Identify the polygon and find the perimeter.

1   (5, 7), (5, 2), (11, 7), (11, 2)

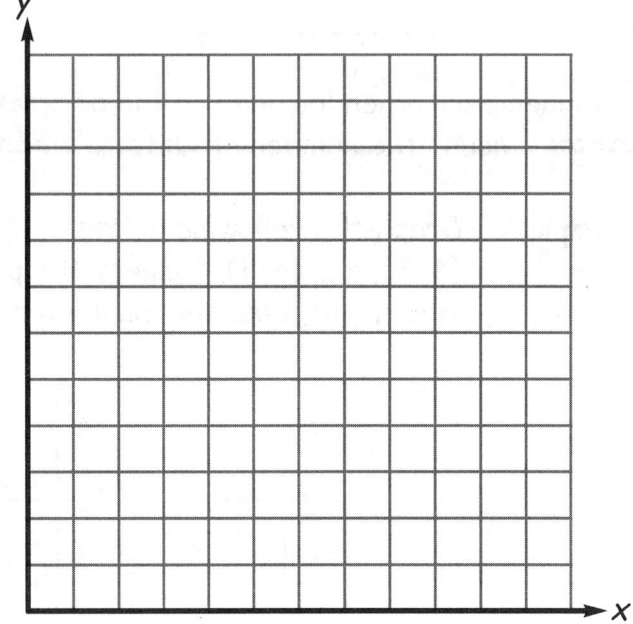

Polygon: _____

Perimeter: _____ units

2   (8, 3), (2, 3), (2, 7), (8, 7)

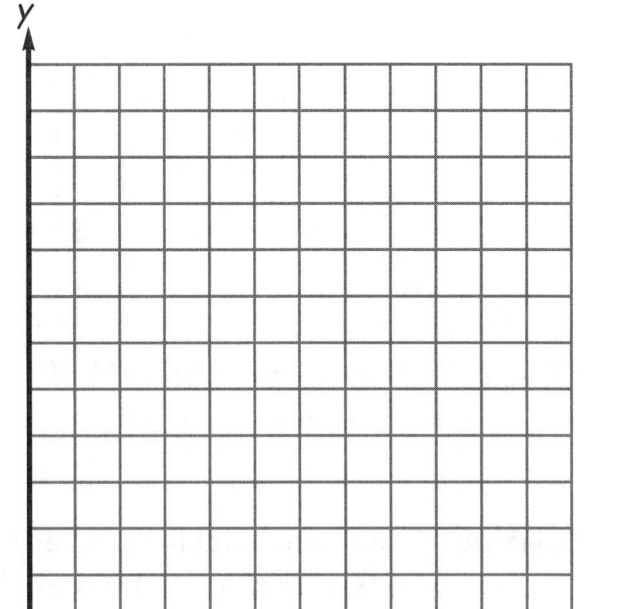

Polygon: _____

Perimeter: _____ units

3 (3, 3), (10, 3), (3, 10), (10, 10)

Polygon: _____

Perimeter: _____ units

TEST PREP

1 What are the coordinates of a point in the coordinate plane that is 3 units to the right of the origin and 4 units up?

A (4, 3)

B (3, 3)

C (3, 4)

D (4, 4)

Directions: Use the graph below to answer questions 2 through 4 in the next column.

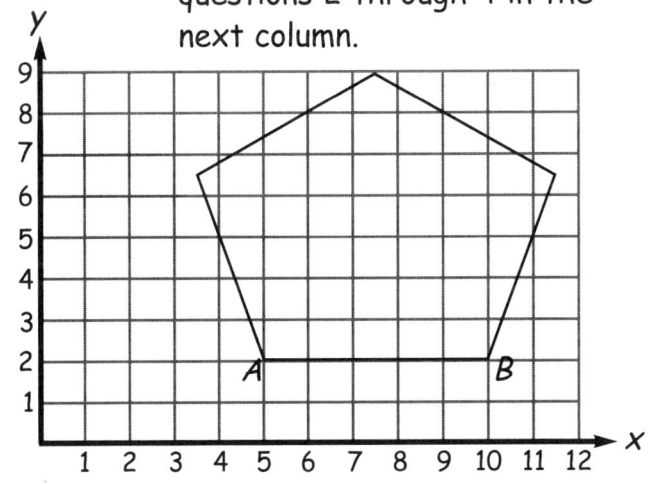

2 Identify the polygon graphed above.

F triangle

G quadrilateral

H pentagon

J hexagon

3 What are the coordinates for point *B*?

A (2, 5)

B (2, 10)

C (5, 2)

D (10, 2)

4 What is the length of side *AB*.

F 2 units

G 8 units

H 5 units

J 10 units

5 Anthony graphed these points: (4, 2), (4, 5), and (6, 2). Which point must be graphed to complete a rectangle

A (4, 6)   **B** (4, 5)

C (6, 4)   **D** (6, 5)

6 What kind of polygon is formed when these points are graphed and connected in order?

(2, 1), (7, 1), (5, 5)

F triangle

G quadrilateral

H pentagon

J hexagon

7 The *x* and *y*-axes divide the coordinate plane into _____

A one quadrant

B three quadrants

C two quadrants

D four quadrants

8 What is the total perimeter, of the two rectangles below? (in units)

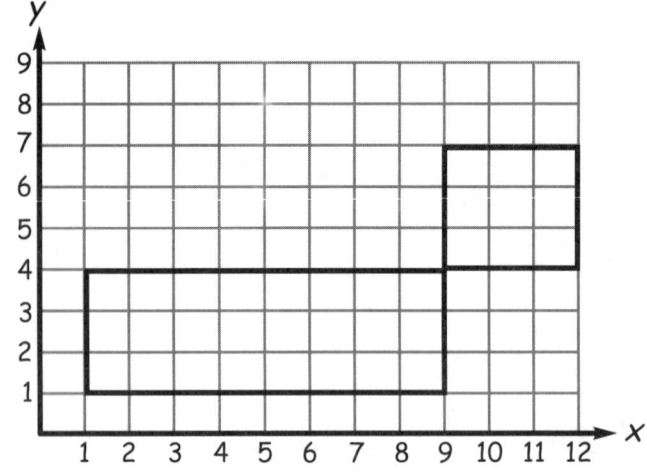

F 34 units

G 22 units

H 16 units

J 12 units

9 **Part A**: Graph points A(4, 3), B(9, 3), C(9, 10), and D(4, 10).

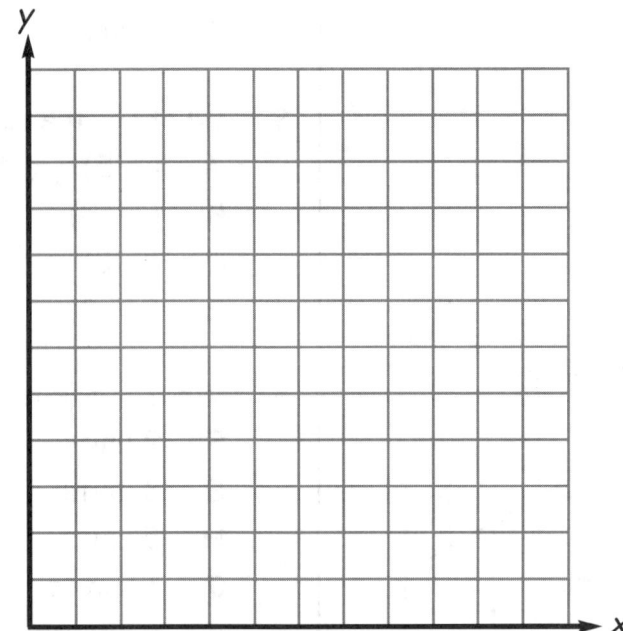

Part B: What type of quadrilateral is formed? _____

9 (continued)

Part C: Find the perimeter of ABCD.

Show your work.

Answer _____

10   Classify the polygon graphed below, and identify the coordinates of each the vertices.

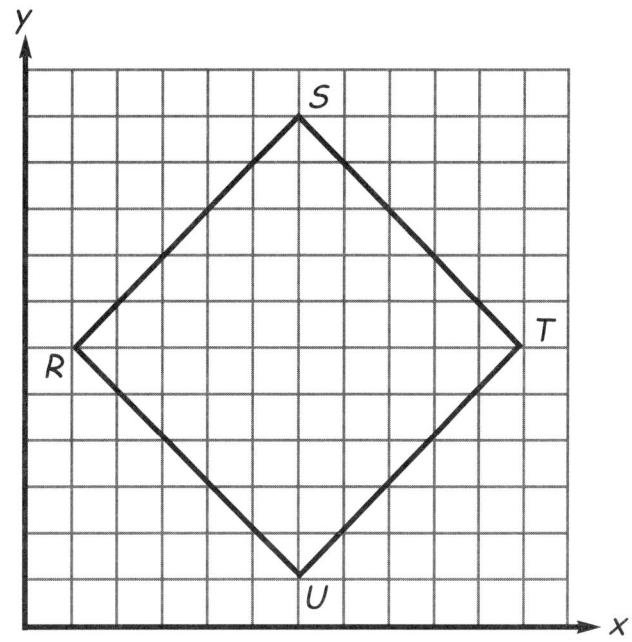

Classification: _____

R coordinates: _____

S coordinates: _____

T coordinates: _____

U coordinates: _____

LESSON SIX

AREA AND VOLUME

6
X-treme

Vocabulary

These words and phrases are associated with Area and Volume and may be used when answering questions in this chapter. Definitions can be found in the Glossary/Index at the back of this *X-treme Review*.

area	irregular polygon	regular polygon
develop formulas	length	volume
height	rectangular prism	width

El Alamillo Bridge
Seville, Spain

What Makes a Strong Bridge?

Bridges are made up of many different geometric figures. One of the shapes most commonly used in the construction of bridges around the world is the triangle. This shape is used most often because it is believed to be the strongest shape, and strength is what you want when building an important route of transportation over a body of water.

In this lesson you will review the classification of geometric shapes and figures. This knowledge will then be used to review the process of calculating the area of regular and irregular shapes, as well as the volume.

When the **Alex Fraser Bridge** opened in 1986, it was the longest cable-stayed bridge (3,050 feet) in the world with a tower height of 505 feet.

Alex Fraser Bridge
Vancover, BC, Canada

Golden Gate Bridge
San Francisco, CA, USA

The Golden Gate Bridge was completed in 1937 – ahead of schedule and under budget – after more than four years of construction at a cost of $35 million. Its suspension span is 4,200 feet long. The bridge's two towers rise 746 feet making them 191 feet taller than the Washington Monument.

El Alamillo bridge opened in time for Spain's hosting of the World Expo in 1992. It spanned the Guadalquivir River in Seville at the north end of La Cartuja island on which the Expo was held. With a 142 meter high pylon, it became a landmark visible from Seville's old town.

Sydney Harbour Bridge opened in 1932 and is the world's largest (but not longest) steel arch bridge and has become a renowned international symbol of Australia. It is 1149 meters long and its arch spans 503 meters. The top of the arch is 134 meters above sea level. The 49 meter wide deck makes Sydney Harbour Bridge the widest longspan bridge in the world.

Sydney Harbour Bridge
Sydney, Australia

http://www.geocities.com/big_bridges

LESSON 6: AREA AND VOLUME

6.1 AREA

The measure of the interior surface of a closed region, is called the (1) _____.
Area is measured in square units.

Directions: Use the following formulas to find the area of triangles and quadrilaterals

Name of Polygon	Formula	Picture
Triangle	$A = \frac{1}{2}bh$	
Square	$A = s \cdot s$ or $A = s^2$	
Rectangle	$A = bh$ or $A = lw$	
Parallelogram	$A = bh$ or $A = lw$	
Rhombus	$A = bh$ or $A = lw$	
Trapezoid	$A = \frac{1}{2}h(b_1 + b_2)$	

Directions: Find the area of the following figures:

Example 1:

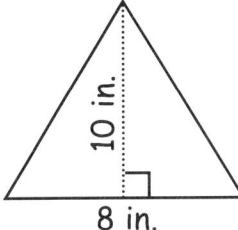

$A = \frac{1}{2}bh$ Write the formula.

$A = \frac{1}{2} \cdot 10 \cdot 8$ Substitute the values for the base and height.

$A = 5 \cdot 8$ Simplify.

$A = 40 \text{ in.}^2$ Square the units.

Example 2:

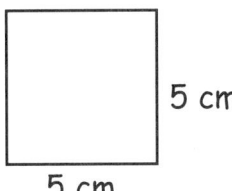

$A = s \cdot s$ Write the formula.

$A = 5 \cdot 5$ Substitute the values for the sides.

$A = 25 \text{ cm}^2$ Simplify and square the units.

Example 3:

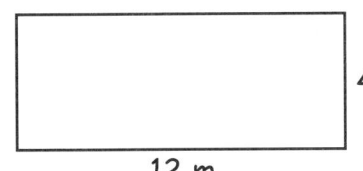

$A = bh$ Write the formula.

$A = 12 \cdot 4$ Substitute the values for the base and height.

$A = 48 \text{ m}^2$ Simplify and square the units.

Example 4:

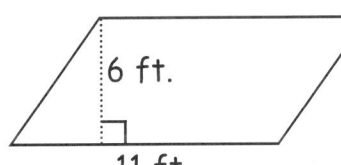

$A = bh$ Write the formula.

$A = 11 \cdot 6$ Substitute the values for the base and height.

$A = 66 \text{ ft.}^2$ Simplify and square the units.

Example 5:

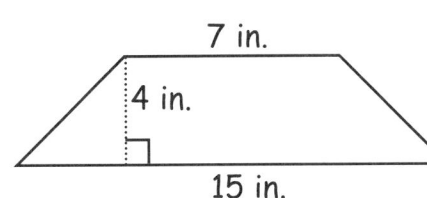

$A = \frac{1}{2}h(b_1 + b_2)$ Write the formula.

$A = \frac{1}{2} \cdot 4(7 + 15)$ Substitute the values for the two bases, and the height.

$A = 2(22)$ Simplify.

$A = 44 \text{ in.}^2$ Square the units.

PRACTICE

Directions: Find the area of the following figures. **Show your work.**

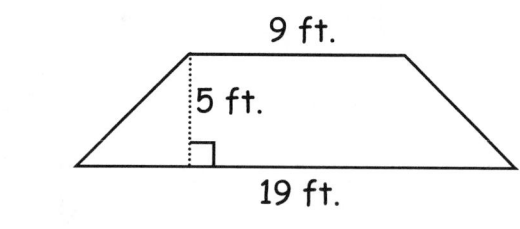

1 5 cm, 15 cm

_____ cm²

2

9 ft., 5 ft., 19 ft.

_____ ft.² (feet squared)

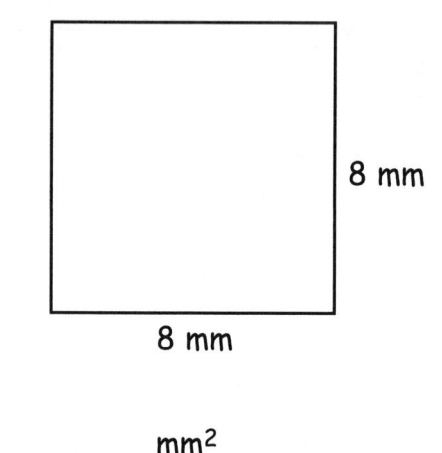

3 8 mm, 8 mm

_____ mm²

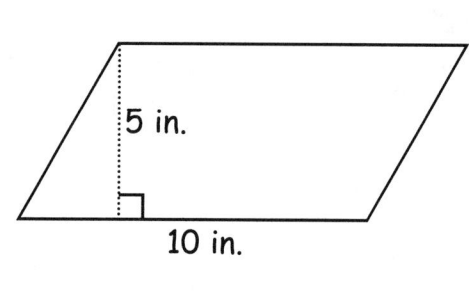

4 5 in., 10 in.

_____ in.²

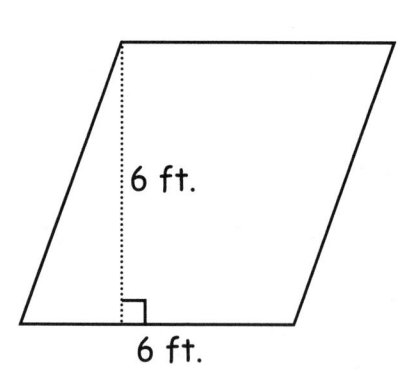

5 6 ft., 6 ft.

_____ ft.²

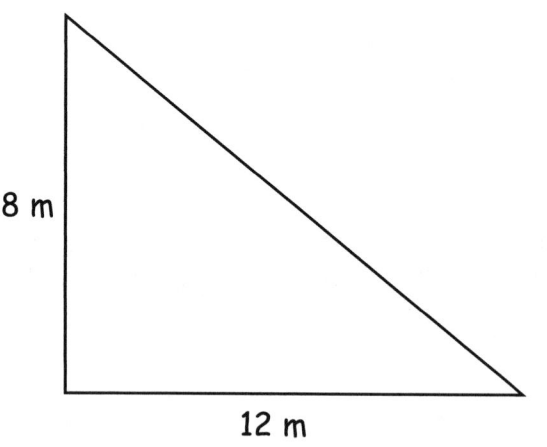

6 8 m, 12 m

_____ m² (square meters)

6.2 REGULAR AND IRREGULAR POLYGONS

Polygons come in all different shapes and sizes. A (1) _____ is a polygon in which all sides and all angles are congruent. An (2) _____ is a polygon whose sides and angles are not all congruent.

Here are some examples of regular and irregular polygons.

Name of Polygon	Regular Polygon	Irregular Polygon
Triangle		
Quadrilateral		
Pentagon		
Hexagon		

Remember: When trying to find the area of an irregular polygon, separate the figure into simple shapes and use the formulas that you already know.

Area of a parallelogram – $A = bh$

Area of a triangle – $A = \frac{1}{2}bh$

Area of a trapezoid – $A = \frac{1}{2}h(b_1 + b_2)$

Example 1: Find the area of the following figure.

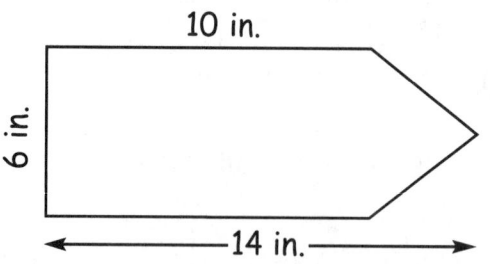

Step 1 Draw a dotted line to separate the figure into "simple" shapes.

Step 2 Separate the figure and re-draw into the two "simple" shapes.

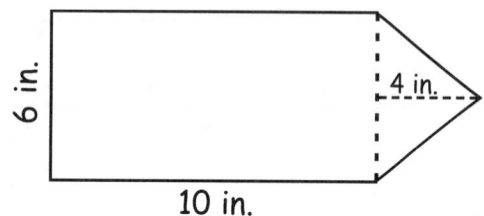 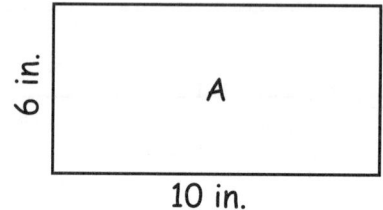 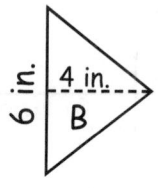

Step 3 Find the area of each part.

Shape A:

$A = bh$

$A = 6 \cdot 10$

$A = 60$ in.^{2}

Shape B:

$A = \frac{1}{2} bh$

$A = \frac{1}{2} \cdot 6 \cdot 4$

$A = 3 \cdot 4$

$A = 12$ in.^{2}

Step 4 Find the total area for the original figure.

Total Area = Area of Shape A + Area of Shape B
Total Area = 60 in.^{2} + 12 in.^{2}
Total Area = 72 in.^{2}

Hint: You can find the area of irregular polygons by adding two areas or subtracting two areas.

Example 2: (Method One) Find the area of an irregular polygon.
   Step 1 Draw a dotted line to separate the figure into "simple" shapes.
   Step 2 Separate the figure and re-draw into the two "simple" shapes.

Irregular Polygon   **Step 1 figure**   **Step 2 figure**

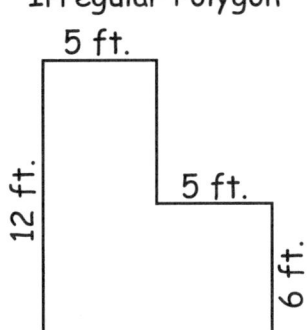 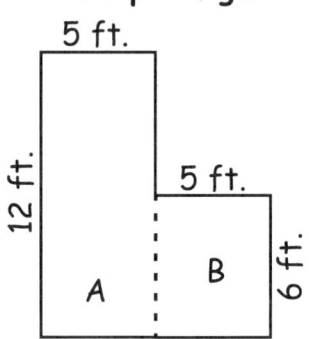 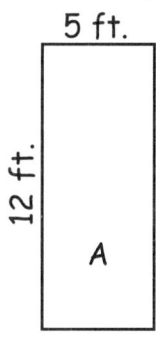

Step 3 Find the area of each part.

Shape A:

$A = bh$

$A = 5 \cdot 12$

$A = 60$ ft.^{2}

Shape B:

$A = \frac{1}{2}bh$

$A = 5 \cdot 6$

$A = 30$ ft.^{2}

Step 4 Find the total area for the original figure.
   Total Area = Area of Shape A + Area of Shape B
   Total Area = 60 ft.^{2} + 30 ft.^{2}
   Total Area = 90 ft.^{2}

OR

Example 2: (Method Two) Fill in the Figure to make a single simple shape and subtract the Parts.

Irregular Polygon   **Step 1 figure**   **Step 2 figure**

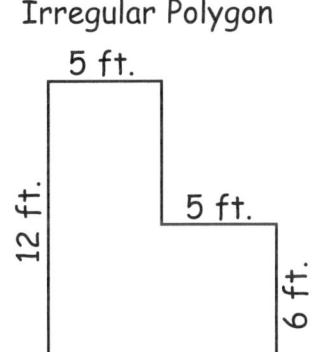 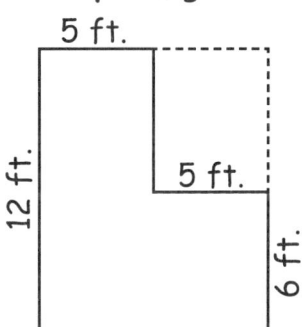 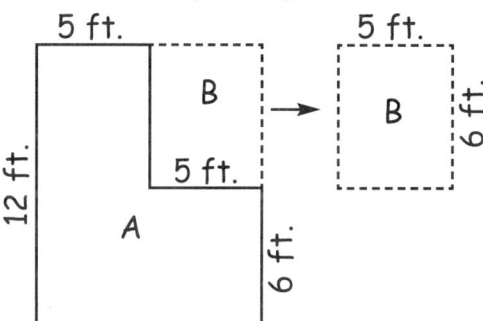

Step 3 Find the area of each part.

Shape A:

$A = bh$

$A = 12 \cdot 10$

$A = 120$ ft.^{2}

Shape B:

$A = bh$

$A = 5 \cdot 6$

$A = 30$ ft.^{2}

Step 4 Find the total area for the original figure.
   Total Area = Area of Shape A – Area of Shape B
   Total Area = 120 ft.^{2} – 30 ft.^{2}
   Total Area = 90 ft.^{2}

PRACTICE

Directions: Find the area of the following figures. Show all of your work.

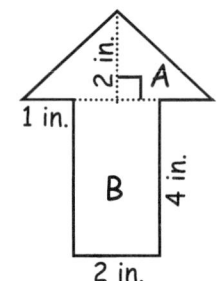

1

_____ inches²

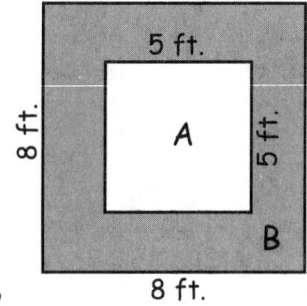

2

_____ square feet

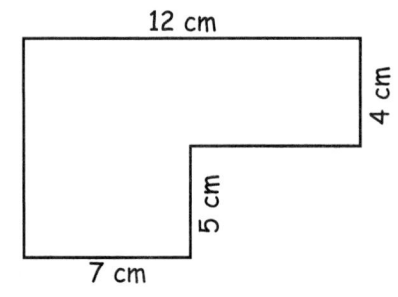

3

_____ centimeters²

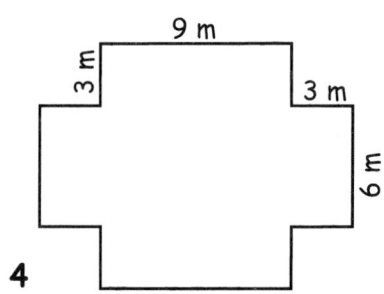

4

_____ square meters

6.3 VOLUME

The number of cubic units needed to fill a solid figure is called the (1) _____.

A (2) _____ is a prism whose six faces are

rectangles. A (3) _____ is a three-dimensional figure (solid) that has

two congruent and parallel faces that are polygons.

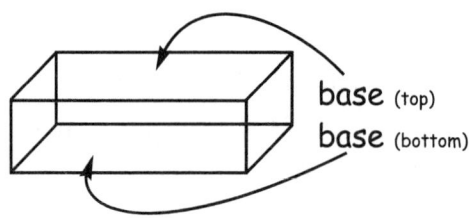

base (top)
base (bottom)

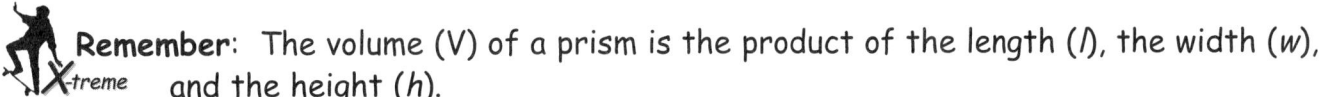 **Remember:** The volume (V) of a prism is the product of the length (*l*), the width (*w*), and the height (*h*).

$$V = lwh$$

Example: Find the volume of the prism.

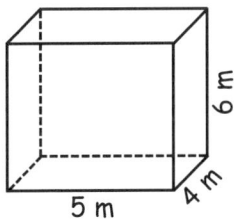

$V = lwh$	Write the formula.
$V = 5 \cdot 4 \cdot 6$	Substitute.
$V = 20 \cdot 6$	Simplify.
$V = 120 \text{ m}^3$	Cube the units.

PRACTICE

Directions: Find the volume of the following rectangular prisms. [not drawn to scale] **Show your work.**

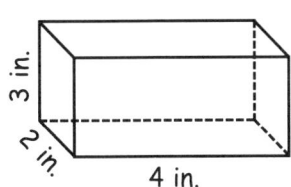

1

_____ cubic inches

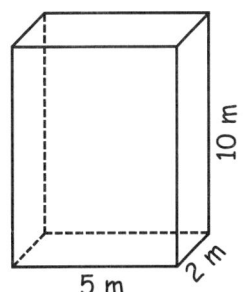

2

_____ meters^3

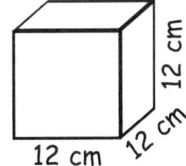

3

_____ centimeters^3

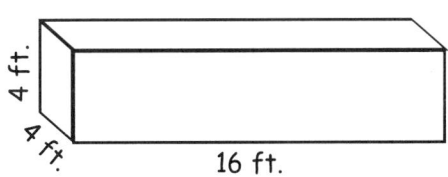

4

_____ cubic feet

TEST PREP

1 What is the volume of the container shown below? [not drawn to scale]

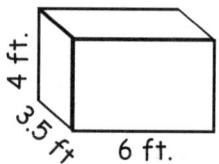

A   13.5 ft.³

B   72 ft.³

C   84 ft.³

D   96.5 ft.³

2 What is the area, in square inches of the triangle below.

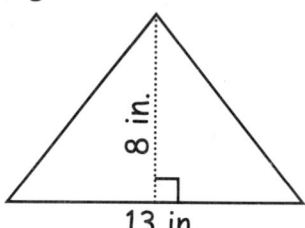

F   26 in.²

G   52 in.²

H   104 in.²

J   208 in.²

3 Find the area of the following trapezoid, in square inches.

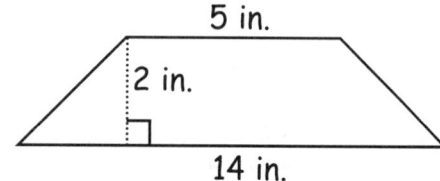

{not drawn to scale]

A   19 in.²

B   24 in.²

C   38 in.²

D   140 in.²

4 Identify the name of the following irregular polygon.

A   quadrilateral

B   pentagon

C   hexagon

D   octagon

5 Find the volume of the cube below, in cubic inches.

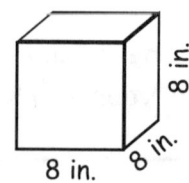

F   24 in.³

G   128 in.³

H   192 in.³

J   512 in.³

6 A polygon with all sides congruent and angles congruent is called a _____.

F   quadrilateral

G   regular polygon

H   rectangle

J   irregular polygon

7 Into what two shapes can the figure below be separated?

A   square and rectangle

B   square and triangle

C   triangle and pentagon

D   triangle and rectangle

8 What is the area of the irregular figure below, in square centimeters.

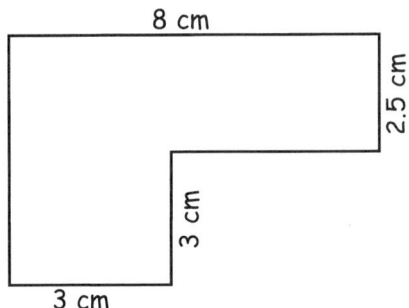

8 cm

2.5 cm

3 cm

3 cm

F   44 cm

G   44 cm²

H   20.5 cm²

J   29 cm²

9 Find the volume of the cube.
Show your work.

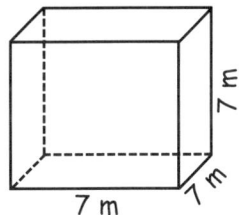

7 m

7 m

7 m

Answer: _____ meters³

10 Find the area of the figure graphed.
Show your work.

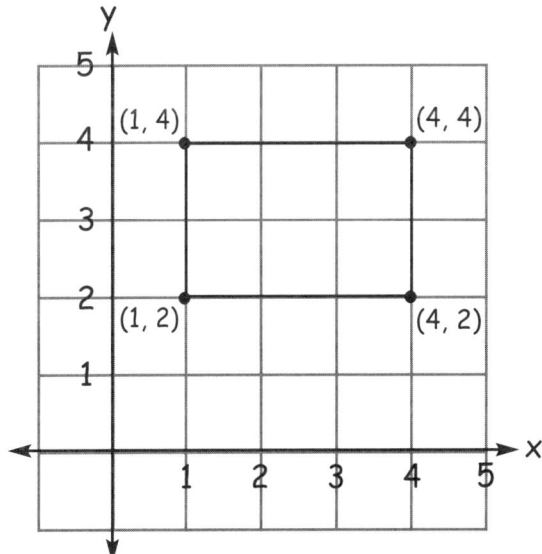

(1, 4)   (4, 4)

(1, 2)   (4, 2)

Answer: _____ units²

11 Find the area of the figure graphed. **Show your work.**

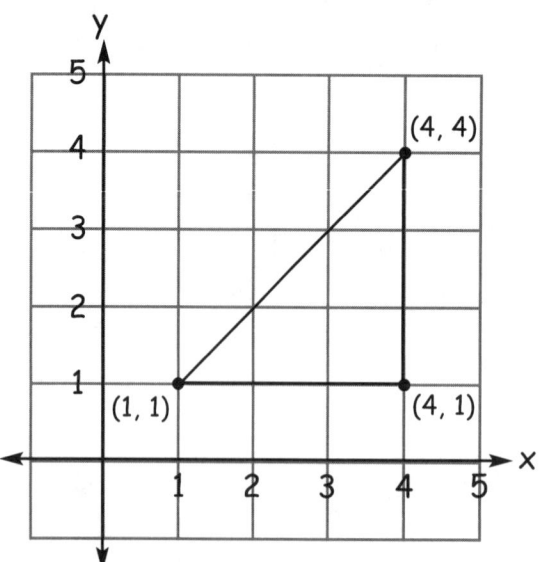

Answer: _____
units²

12 Amanda wants to figure out the area of her grandmother's garden. A diagram of the fenced in area is shown below.

What is the total area, in square centimeters, of the fenced area? [not drawn to scale] **Show your work.**

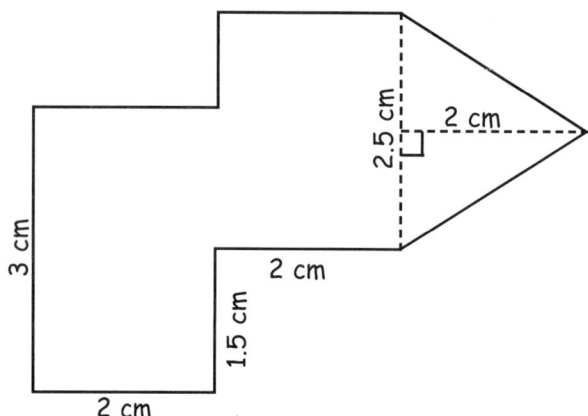

Answer: _____ cm²

LESSON SEVEN

CIRCLE GEOMETRY

7
X-treme

Vocabulary
These words and phrases are associated with Circle Geometry and may be used when answering questions in this chapter. Definitions can be found in the Glossary/Index at the back of this *X-treme Review*.

arc	circumference	radius
chord	diameter	sector

It's a Wonder!

The Great Pyramid, the **Pyramid of Khufu**, is the oldest and only surviving *Seven Wonders of the World*. It was built in Egypt around 2560 B.C. and the full construction took almost 20 years. Starting at 481 feet high when first built, the Great Pyramid lost about 10 feet of its top over the years. Up until the nineteenth century, the Great Pyramid was the tallest structure in the world. The length of each side was 751 ft., pretty big structure.

The Great Sphinx and the Pyramid of Khafra

Another interesting fact that has been proven about the Great Pyramid involves the perimeter of the base. First find the circumference of a circle using the height of the pyramid as the radius:

Original Height	More Recent Height
$C = 2\pi r$	$C = 2\pi r$
$C = 2 \cdot \pi \cdot 481$	$C = 2 \cdot \pi \cdot 471$
$C = 962 \cdot \pi$	$C = 942 \cdot \pi$
$C \approx 3022$ ft.	$C \approx 2959$ ft.

This circumference is said to be equal to the perimeter of the pyramid:

$P = 4s$
$P = 4 \cdot 751$
$P = 3004$ ft.

Wow! Look how close that is! Do you think it was done on purpose?

In this lesson, you will review all the parts of a circle and how to use some of those parts to determine the circumference and area.

http://ce.eng.usf.edu/pharos/wonders/pyramid.html
http://www.greatbuildings.com/
http://gei.aerobatics.ws/egypt_great_pyramids.html

LESSON 7: CIRCLE GEOMETRY

7.1 CIRCLES

The (1) _____ of a circle is a line segment that extends from the center of a circle to any point on the circle's perimeter.

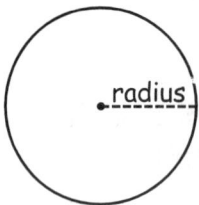

Remember: To find the radius when given the diameter, divide the diameter by two.

Example 1: Identify the radius of the circles below.

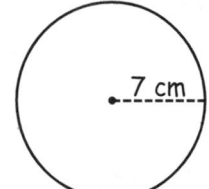

The radius is 7 cm.

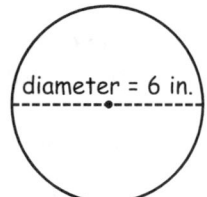

The radius is 3 in.

The (2) _____ of a circle is a (3) _____ that passes through the center of the circle.

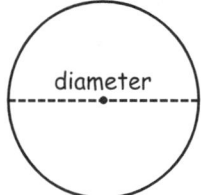

Remember: The diameter of a circle is twice the length of the radius. So, to find the diameter, when given the radius, multiply the radius by two.

Example 1: Identify the diameter of the circles below.

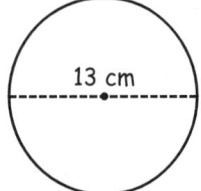

The diameter is 13 cm.

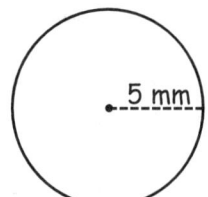

The diameter is 10 mm.

PRACTICE

Directions: Find the length of the <u>radius</u> for each circle.

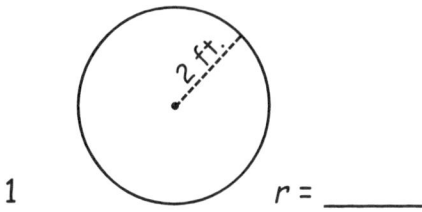

1    $r =$ _____

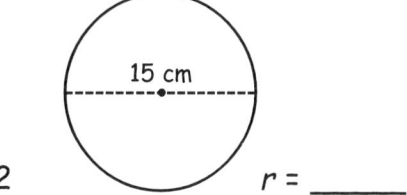

2    $r =$ _____

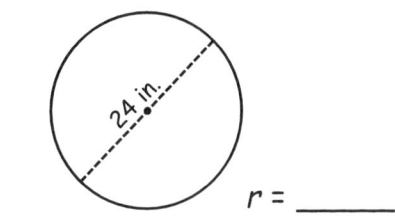

3    $r =$ _____

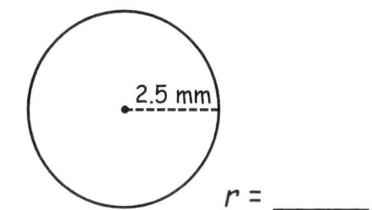

4    $r =$ _____

Directions: Find the length of the <u>diameter</u> for each circle.

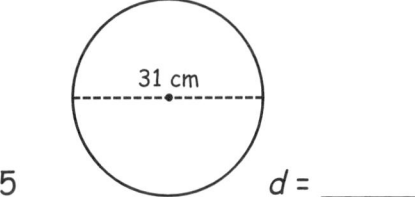

5    $d =$ _____

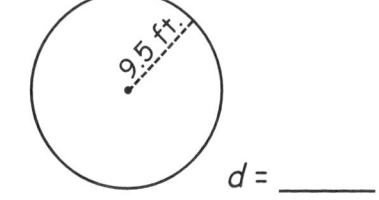

6    $d =$ _____

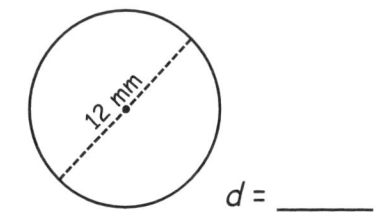

7    $d =$ _____

8    $d =$ _____

7.2 Circumference and Area

The (1) _____ of a circle is the distance around the circle.

Remember: The ratio of every circle's circumference to its diameter is the same, π. You can use 3.14159 as the approximation for π.

$$C = \pi d \qquad C = 2\pi r$$

Example 1: Find the circumference of the circle. Round your answer to the nearest hundredth (two places to the right of the decimal).

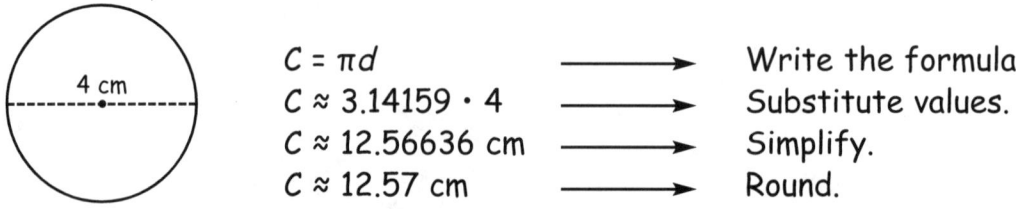

$C = \pi d$ $\longrightarrow$ Write the formula.
$C \approx 3.14159 \cdot 4$ $\longrightarrow$ Substitute values.
$C \approx 12.56636$ cm $\longrightarrow$ Simplify.
$C \approx 12.57$ cm $\longrightarrow$ Round.

Example 2:

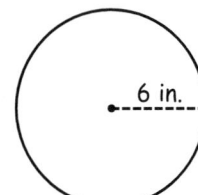

$C = 2\pi r$ $\longrightarrow$ Write the formula.
$C \approx 2 \cdot 3.14159 \cdot 6$ $\longrightarrow$ Substitute values.
$C \approx 6.28318 \cdot 6$ $\longrightarrow$ Simplify.
$C \approx 37.69908$ in.
$C \approx 37.70$ in. $\longrightarrow$ Round.

Remember: The area of a circle equals the product of π and the square of the radius.

$$A = \pi r^2$$

Example 1: Find the area of the circle. Round your answer to the nearest hundredth.

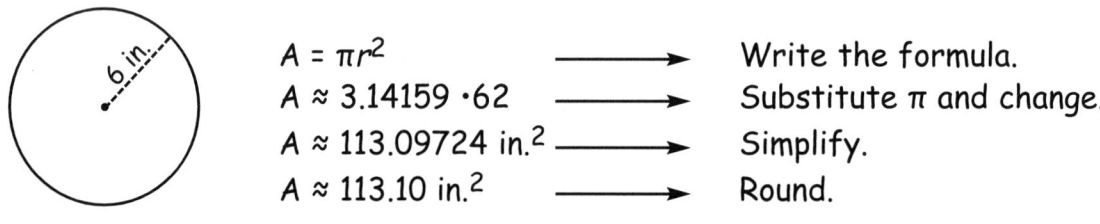

$A = \pi r^2$ $\longrightarrow$ Write the formula.
$A \approx 3.14159 \cdot 6^2$ $\longrightarrow$ Substitute π and change.
$A \approx 113.09724$ in.^{2} $\longrightarrow$ Simplify.
$A \approx 113.10$ in.^{2} $\longrightarrow$ Round.

Example 2:

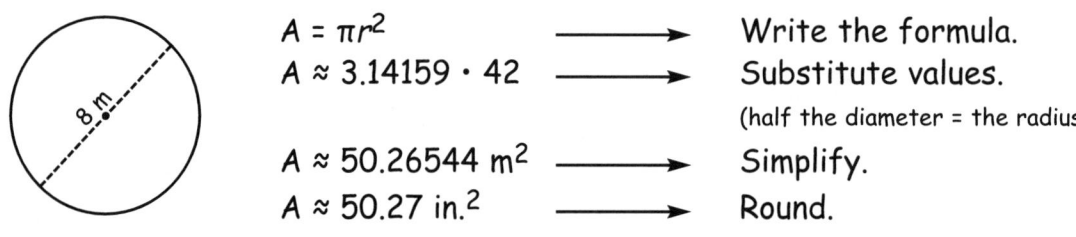

$A = \pi r^2$ $\longrightarrow$ Write the formula.
$A \approx 3.14159 \cdot 4^2$ $\longrightarrow$ Substitute values.
(half the diameter = the radius)
$A \approx 50.26544$ m^2 $\longrightarrow$ Simplify.
$A \approx 50.27$ in.^{2} $\longrightarrow$ Round.

PRACTICE

Directions: Find the circumference of each circle where *r* = radius and *d* = diameter. Use 3.14159 for π.

1 *d* = 4 cm

2 *r* = 33 m

3 *r* = 8 in.

4 *d* = 34 mm

5 *d* = 10 yd.

6 *r* = 40 ft.

Directions: Find the area of each circle where *r* = radius and *d* = diameter. Use 3.14159 for π. Round your answers to the nearest hundredth.

7 *r* = 7 in.

8 *d* = 22 cm

9 *d* = 9 yd.

10 *r* = 100 km

11 *r* = 5 mi.

12 *d* = 40 mm

🏃.3 Area of Sectors

The part of a curve between any two of its points is called an (1) _____.

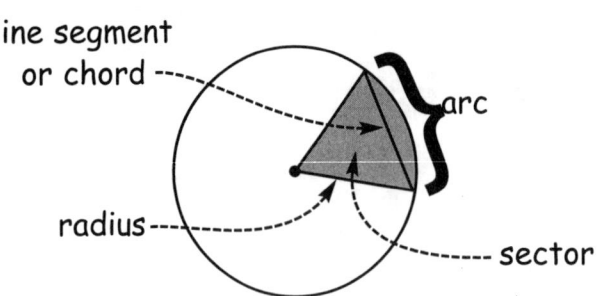

line segment
or chord ---
arc
radius ---
sector

The region of the circle formed by two radii and their intercepted arc is called the
(2) _____ of a circle.

Remember: To find the area of a sector of a circle, it is necessary to measure the
central angle (n) and the length of the radius.

$$A = \frac{n}{360°} \pi r^2$$

Example 1: Find the area of the sector if the central angle = 24°, and the radius = 4 in.
Use 3.14159 for π. Round your answer to the nearest hundredth.

$A = \dfrac{n}{360°} \pi r^2$ ⟶ Write the formula.

$A \approx \dfrac{24°}{360°} \cdot 3.14159 \cdot 4^2$ ⟶ Substitute.

$A \approx 0.2094393 \cdot 16$ ⟶ Simplify.

$A \approx 3.3510588$ in.^{2} ⟶

$A \approx 3.35$ in.^{2} ⟶ Round.

Example 2:

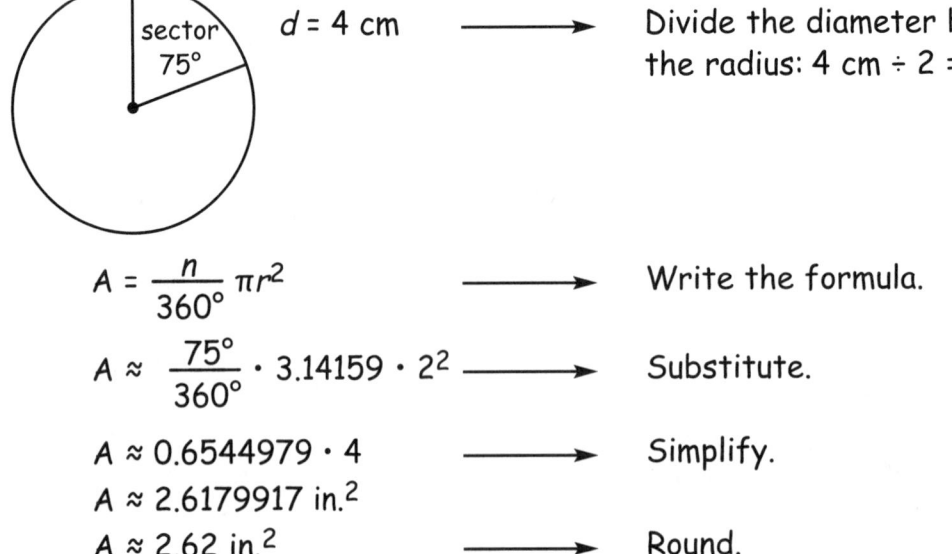

sector
75°

$d = 4$ cm ⟶ Divide the diameter by 2 to find
the radius: 4 cm ÷ 2 = 2 cm.

$A = \dfrac{n}{360°} \pi r^2$ ⟶ Write the formula.

$A \approx \dfrac{75°}{360°} \cdot 3.14159 \cdot 2^2$ ⟶ Substitute.

$A \approx 0.6544979 \cdot 4$ ⟶ Simplify.

$A \approx 2.6179917$ in.^{2}

$A \approx 2.62$ in.^{2} ⟶ Round.

PRACTICE

Directions: Find the area of each sector where *n* = the measure of the central angle, *r* = radius, and *d* = diameter. Use 3.14159 for π. Round your answers to the nearest hundredth.

1   *n* = 42°, *r* = 12 mm

2   *n* = 180°, *r* = 5 ft.

3   *n* = 90°, *d* = 18 cm

4   *n* = 37°, *d* = 31 in.

TEST PREP

1   Identify the part of the circle labeled *a* in the diagram below.

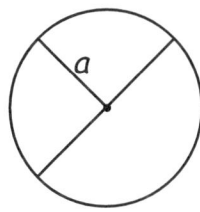

    A   arc
    B   chord
    C   diameter
    D   radius

2   Find the area of a sector when the central angle is 60° and the diameter is 8 m.
    F   8.38 m²
    G   838 m²
    H   33.51 m²
    J   3351 m²

3   What is the circumference of the circle below?

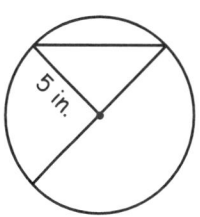

    A   15.7 in.
    B   31.4 in.
    C   78.5 in.
    D   314 in.

4   What is the circumference of a circle whose diameter is 17 ft.? Use 3.14159 for π.
    F   26.71 ft.
    G   53.38 ft.
    H   53.41 ft.
    J   106.81 ft.

5 What is the area of the circle?

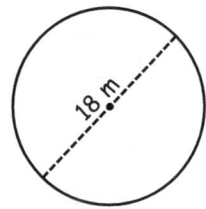

18 m

A   56.52 m²
B   254.47 m²
C   324 m²
D   1017.36 m²

6 What is the area of the sector in the circle below if the radius is 9 cm. Round to the nearest hundredth.

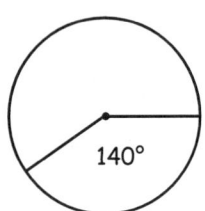

140°

F   49.00 cm²
G   49.50 cm²
H   98.96 cm²
J   99.00 cm²

7 In the following diagram, identify the arc:

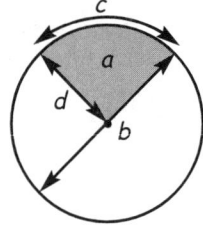

A   *a*
B   *b*
C   *c*
D   *d*

8 Find the circumference of a circle whose radius is 5 in. Use 3.14159 for π, and round to the nearest tenth.

F   15.7 in.
G   31.4 in.
H   22.5 in.
J   47.1 in.

9 The Milanese family orders pizza for dinner. Each pie has a diameter of 16 inches. If one pizza is divided into eight sections, what is the area of each section? Round your answers to the nearest whole number.
Show your work.

Answer: _____ in.²

10 The diameter of a circular pool is 32 ft. What is the circumference of the pool? Round your answers to the nearest tenth. **Show your work.**

Answer: _____ ft.

11 Find the area of the sector below. Round your answers to the nearest hundredth. **Show your work.**

90°

42 mm

Answer: _____ mm^2

12 If the radius of a circle is double, what happens to the original circumference? Illustrate and justify your answer.

_____

_____

_____

_____

LESSON EIGHT

MEASUREMENT

Vocabulary

These words and phrases are associated with Measurement and may be used when answering questions in this chapter. Definitions can be found in the Glossary/Index at the back of this *X-treme Review*.

capacity	metric system
customary measurement system	metric units (gram, liter, meter)
customary unit	rectangular prism

Don't Try This At Home!

For years one of the most common bets was: I bet you can't drink a gallon of milk in one hour!

Many have tried and some succeeded, but not for that long. Unfortunately, the milk usually comes back out! Scientifically speaking, the main reason that it is not possible to drink a gallon of milk in an hour is related to how much the stomach can hold, which is the stomach's capacity. The capacity of the stomach is about 1 liter which is about $1/5$ of a gallon. When stomach capacity is exceeded, the natural response will be for you to get sick. So, don't try this at home. There is a very good chance you will not make it through even half of the gallon.

This lesson will review the customary and metric units of capacity. Also, you will review how to measure the capacity and find the volume of rectangular prisms.

http://www.madsci.org/posts/archives/2001-04/988043127.An.r.html
http://hemsidor.torget.se/users/b/bohjohan/convert/conv_e.htm
http://graphics.samsclub.com/images/products/0004190002763_LG.jpg

http://www.lsuagcenter.com/en/communications/leads/Louisianas+Got+Milk+Scientists+Keep+Dairy+Industry+Alive.htm

LESSON 8: MEASUREMENT

7.1 METRIC UNITS OF CAPACITY

The (1) _____ is a system of measurement based on the decimal system; the standard unit of length is a meter, of capacity is a liter, and of mass is a gram.

(2) _____ are the units used in the metric system.

(3) _____ is the maximum amount a container can hold.

Metric Units of Capacity

Kiloliter	Hectoliter	Decaliter (Dekaliter)	Liter	Deciliter	Centiliter	Milliliter
1000 L	100 L	10 L	1 L	0.1 L	0.01 L	0.001 L

$$1 \text{ L} = 1000 \text{ mL}$$

Remember: To rename a larger unit of capacity with a smaller unit, multiply by a power of 10.

Example 1: 9.1 kL = _____ L

$$\underline{9.1 \text{ kL} \cdot 1000} = 9100 \text{ L}$$
$$\downarrow$$
$$1 \text{ kL} = 1000 \text{ L}$$

Remember: To rename a smaller unit of capacity with a larger unit, divide by a power of 10.

Example 2: 2500 mL = _____ L

$$\underline{2500 \text{ mL} \div 1000} = 2.5 \text{ L}$$
$$\downarrow$$
$$1000 \text{ mL} = 1 \text{ L}$$

PRACTICE

Directions: Determine the equivalent metric unit of capacity.

1   3 L = _____ mL

2   0.7 kL = _____ L

3   40000 L = _____ kL

4   7000 mL = _____ L

5   530 hL = _____ L

6   7 kL = _____ mL

Directions: Circle the letter of the most likely measure.

7   the capacity of a small thermos

   a   0.5 kL

   b   5000 mL

   c   0.5 L

8   the capacity of a teacup

   a   2.4 mL

   b   240 mL

   c   2.4 L

9   the capacity of a teaspoon

   a   0.05 kL

   b   0.05 L

   c   5 mL

8.2 CUSTOMARY UNITS OF CAPACITY

The (1) _____ is the system of measurement used mainly in the United States to measure length , mass, and capacity.

(2) _____ are the units of measure used in the customary measurement system.

> 1 cup (c.) = 8 fluid ounces (fl. oz.)
> 1 pint (pt.) = 2 c.
> 1 quart (at.) = 2 pts.
> 1 gallon (gal.) = 4 qts.

Remember: To rename a larger unit with a smaller unit, multiply.

X-treme

Example 1:   2 qts. = _____ c.

$$\underline{2 \text{ qts.} \cdot 4} = 8 \text{ cups}$$

1 qt. = 2 pts.
and 1 pt. = 2 c.

2 qts. = 8 cups

Remember: To rename a smaller unit with a larger unit, divide.

X-treme

Example 2:   12 fl. ozs. = _____ c.

$$\underline{12 \text{ fl. ozs.} \div 8} = 1.5 \text{ fl. ozs.}$$

1 c. = 8 fl. ozs.

PRACTICE

Directions: Determine the equivalent customary unit of capacity.

1   5 gal = _____ qt.   **2**   8 pt = _____ c.   **3**   3.5 qt = _____ pt.

4   16 fl oz = _____ pt.   **5**   6 c = _____ qt.   **6**   2 gal = _____ pt.

Directions: Circle the letter of the most appropriate measure.

7   a can of grape juice

    a   48 pt.   **b**   48 qt.   **c**   48 fl. ozs.

8   the capacity of a canteen

    a   1 gal.   **b**   1 qt.   **c**   1 fl oz.

8.3 MEASUREMENT AND PRECISION OF CAPACITY

Remember: The smaller the unit of measurement used, the more precise the measurement is.

Example: When two measures are given for the same capacity, such as 1 pint equals 2 cups, it is usually understood that 2 cups is the *more* precise measurement, because it is precise to the nearest cup.

PRACTICE

Directions: Circle the <u>more</u> <u>precise</u> measurement.

1   16 ozs. or 1 lb.

2   180 mL or 18 cL

3   4 qt. or 1 gal.

4   1 qt. or 2 pts.

5   0.24 L or 240 mL

8.4 VOLUME OF A RECTANGULAR PRISM

A (1) _____ is a three-dimensional figure (solid) that has two congruent and parallel faces that are polygons.

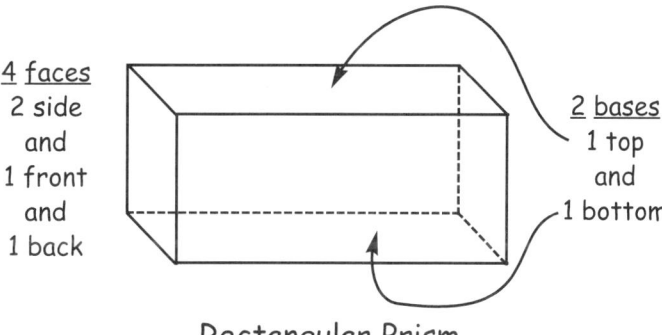

<u>4 faces</u>
2 side
and
1 front
and
1 back

<u>2 bases</u>
1 top
and
1 bottom

Rectangular Prism

Remember: The volume of a prism is the product of the length (*l*), the width (*w*), and the height (*h*).

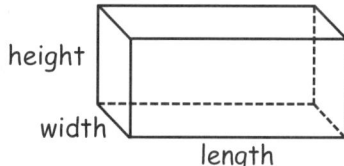

height

width

length

$V = lwh$

Example: Find the volume of the prism.

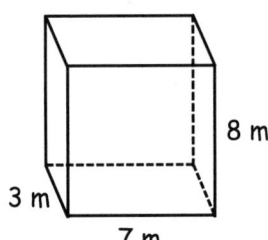

$V = lwh$ ⟶ Write the formula.
$V = 7 \cdot 3 \cdot 8$ ⟶ Substitute.
$V = 21 \cdot 8$ ⟶ Simplify.
$V = 168 \text{ m}^3$ ⟶ Cube the units.

PRACTICE

Directions: Find the volume of each rectangular prism.

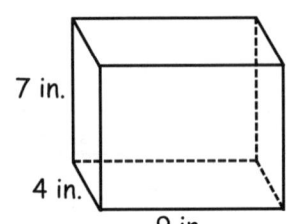

1   _____ in.^{3}

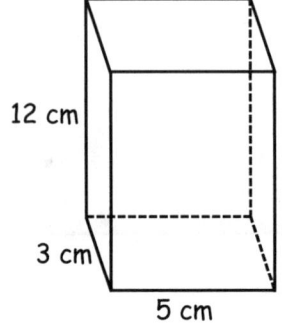

2   _____ cm^3

TEST PREP

1   Which is the largest amount?

A   3 c 12 fl oz

B   73 fl oz

C   1 gal

D   2 pt 1 c

2   Jason drank one fourth of a liter of soda. How many milliliters did he drink?

F   2.5 mL

G   25 mL

H   250 mL

J   2,500 mL

3   What is the volume of the container shown below?

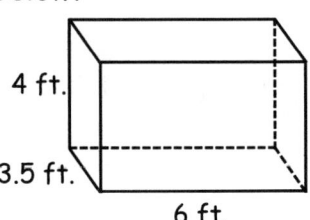

A   13.5 ft.^{3}

B   72 ft.^{3}

C   84 ft.^{3}

D   96.5 ft.^{3}

4   Which is the most likely measure for the capacity of a pitcher?

F   0.6 kL

G   6000 mL

H   0.6 L

J   1600 L

5 Which is the most precise measurement for the capacity of a bowl of soup?

   A   0.3 L

   B   0.32 L

   C   320 mL

   D   3200 mL

6 Eight gallons equals which of the following?

   F   4 pts

   G   4 qts

   H   32 pts

   J   32 qts

7 Julie is dyeing T-shirts purple. She needs 6 cups of dye for each T-shirt. If she wants to dye 3 T-shirts, how many pints of dye does she need?

   A   18 pts

   B   9 pts

   C   6 pts

   D   2 pts

8 5.2 L equals which of the following?

   F   5200 mL

   G   520 mL

   H   0.52 mL

   J   0.0052 mL

9 Find the volume of the rectangular prism when $x = 8$ inches [not drawn to scale].

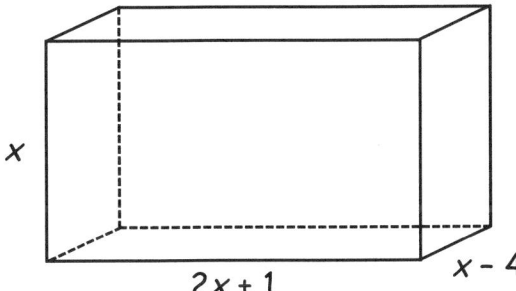

Answer: _____ in.^{3}

10 Jan needed 96 fl oz of lemon juice to make lemonade for a party. How many cups of lemon juice is this?

Show your work.

Answer: _____ cups

11 Raul blends a drink that has 1 egg, 0.25 L of milk, and 0.25 L of orange juice. To make 7 of these drinks, how many milliliters of orange juice does he need?

Show your work.

Answer: _____ mL

12 Mr. Smith wants to build a storage box that will hold 400 ft.^{3} of salt. If the box has a square base, 5 ft. by 5 ft., how high must the box be?

Show your work.

Answer: _____ ft.

Vocabulary

These words and phrases are associated with Statistics and Probability and may be used when answering questions in this chapter. Definitions can be found in the Glossary/Index at the back of this *X-treme Review*.

bar graph	line graph	mode	probability
event	mean	outcome	range
interval	median	predict	

How Many People Are Here?

It is very important to be able to make predictions. Now, we are talking about predictions made based on years of data, not a feeling someone has. Population growth or decline is something that is predicted for many reasons such as: jobs, housing, water supply, pollution, and many more. It is important for cities and states to plan ahead. The graph at the right is an actual prediction of the population growth in New York State for the next 24 years. What do you think the prediction will look like for 2040?

In this lesson, you will review how to determine the probability of an event occurring. You will also review how to make predictions based on of a set of data.

http://www.npg.org/states/ny.htm

_____

Forecast of World Population

The future growth of population is difficult to predict. Birth rates are declining slightly on average, but vary greatly between developed countries (where birth rates are often at or below replacement levels) and developing countries. Death rates can change unexpectedly due to disease, wars and catastrophes, or advances in medicine. Over the last 10 years, the UN has consistently revised these projections downward.

Current projections by the UN's Population Division, based on the 2004 revision of the World Population Prospects database [3], are as follows.

Year	2010	2020	2030	2040	2050
Population (billions)	6.8	7.6	8.2	8.7	8.9

http://en.wikipedia.org/wiki/World_population

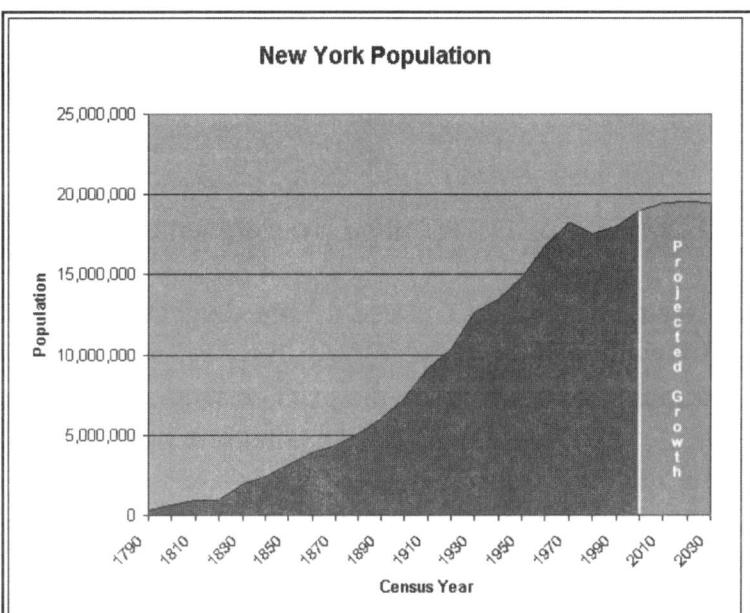

Year	Population		Year	Population		Year	Population
1790	340,120		1880	5,082,871		1970	18,241,391
1800	589,051		1890	6,003,174		1980	17,558,165
1810	959,049		1900	7,268,894		1990	17,990,455
1820	1,372,812		1910	9,113,614		2000	18,976,457
1830	1,918,608		1920	10,385,227		2010	19,443,672
1840	2,428,921		1930	12,588,066		2020	19,576,920
1850	3,097,394		1940	13,479,142		2030	19,477,429
1860	3,880,735		1950	14,830,192			
1870	4,382,759		1960	16,782,304			

LESSON 9: STATISTICS AND PROBABILITY

9.1 PROBABILITY

(1) _____ is the chance of an event occurring.

$$P(event) = \frac{Number\ of\ Favorable\ Outcomes}{Number\ of\ Possible\ Outcomes}$$

(2) An _____ is the possible result of an action.

An (3) _____ is any outcome or group of outcomes.

Example 1: Find P(rolling a 5) on a standard die.

$\frac{1}{6}$ ⟶   # of favorable outcomes

      ⟶   # of possible outcomes

P(rolling a 5) = $\frac{1}{6}$ (or a 1 in 6 possibility of a 5)

Example 2: Using the same die, find P(rolling an even number)

$\frac{3}{6}$ ⟶   # of favorable outcomes

      ⟶   # of possible outcomes

P(rolling an even number) = $\frac{3}{6}$ (or a 3 in 6 possibility of a 2, 4, or 6)

Remember: You can use a tree diagram to display and count possible outcomes.

Example 3: Use a tree diagram to find the sample space for tossing two coins. Find the probability of tossing one heads.

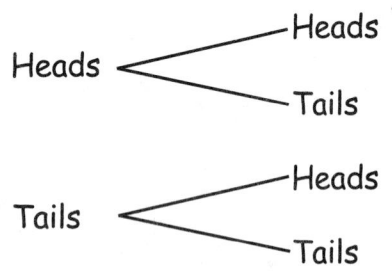

Heads
    Heads
    Tails

Tails
    Heads
    Tails

The tree diagram shows that there are four possible outcomes:

(H, H), (H, T), (T, H), (T, T),

P(one heads) = $\frac{1}{2}$ (or a 1 in 2 possibility of a heads or a tails)

PRACTICE

Directions: Find the probability for each of the following.

1 A fair die is rolled. Find the probability of rolling each of the following:

    a   P (greater than 4)

    b   P (odd number)

    c   P (less than or equal to 6)

2 A bag contains 7 green marbles, 4 blue marbles, 1 yellow marble, 3 white marbles, and 2 red marbles. If a marble is chosen at random, find the probability of each:

    a   P (yellow)

    b   P (black)

    c   P (green)

3 Use a tree diagram to list all of the possible combinations of the letters C, A, and T, without repeating letters.

9.2 MEAN, MEDIAN, MODE, AND RANGE

(1) _____, (2) _____, and (3) _____ are the three most common measures of central tendencies.

The (4) _____, or average is the quotient obtained when the sum of the numbers in a set is divided by the number of data items.

Remember: When determining the mean, add all the numbers then divide by the amount of numbers that were added.

Example: Consider the data 1, 3, 5, 5, 7, 9, and 12.

$$\text{mean} = \frac{1 + 3 + 5 + 5 + 7 + 9 + 12}{7}$$

$$\text{mean} = \frac{42}{7}$$

$$\text{mean} = 6$$

The (1) _____ is the middle number of a set of data arranged in increasing or decreasing order. If there is no middle number, the median is the average of the two middle numbers.

Example 1: Consider the data 5, 6, 9, 3, 2, 3, and 7.

2, 3, 3, 5, 6, 7, 9   ⟵——— Numbers must be in order.

Number must be in order.

Example 2: Consider the data 5, 9, 6, 9, 3, 2, 3, and 7.

2, 3, 3, 5, 6, 7, 9, 9 ⟵——— Numbers must be in order.

Since there is no middle number, take
the average of the two middle numbers.

$$\text{median} = \frac{5 + 6}{2}$$

$$\text{median} = \frac{11}{2}$$

$$\text{median} = 5.5$$

The (2) _____ of a set of data is the number or numbers that occur most frequently.

Example: Consider the data 3, 5 , 6, 6, 7, and 10. The mode is 6 because it occurs most often.

The (3) _____ of a group of data is the difference between the greatest and least values in the set.

Example: Consider the data 23, 24, 24, 25, 26, 27, and 27.

23 is the smallest data value in the set.
27 is the largest data value in the set.

27 – 23 = 4 ⟵——— 4 is the range of the data set.

PRACTICE

Directions: For each data set determine the mean, median, mode, and range,

1   12, 21, 17, 13, 19, 12, 19, 20, 22

    a   Mean: _____

    b   Median: _____

    c   Mode: _____

    d   Range: _____

2   7, 4, 1, 11, 6, 7, 8, 2, 5, 12, 6, 6, 9, 11, 3

    a   Mean: _____

    b   Median: _____

    c   Mode: _____

    d   Range: _____

9.3 READING AND INTERPRETING GRAPHS

Data is often organized and represented visually with graphs. A (1) _____
is a graph that uses horizontal or vertical bars to display data.

Example: To "read" a bar graph, compare the length (top) of the bar with the units identified on the left side of the graph. For example, on Monday, 8 tickets were sold. By adding all of the bars, a total of (8 + 4 + 10 + 1 +15) 38 tickets were sold for the week.

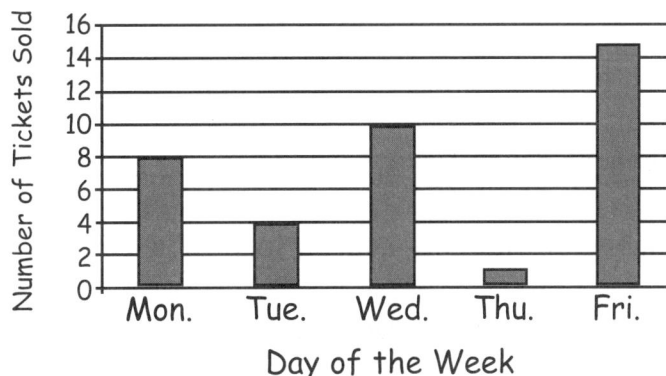

6th Grade Dance – Ticket Sales

A (1) _____ is used to show trends or changes over time.

Example: To "read" a line graph, compare the position of the "data dots" on the graph and connecting lines with the units identified on the left side of the graph. For example, the most number of students, 450, used the library in May and the greatest increase in students using the library was between April and May.

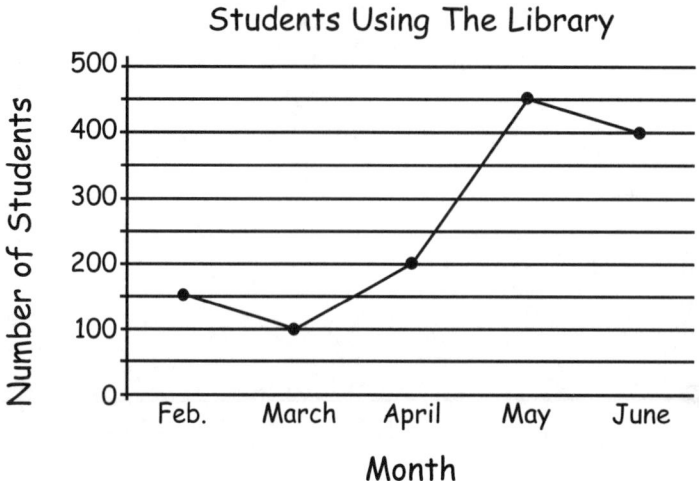

Students Using The Library

PRACTICE

Directions: Use the bar graph below to answer questions 1 – 5.

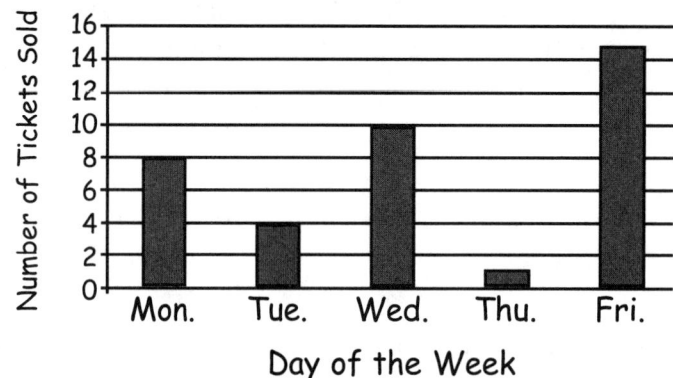

6th Grade Dance – Ticket Sales

1 About how many dance tickets are sold on Thursday? _____

2 On which day are the most dance tickets sold? _____

3 What is the total number of students coming to the dance? _____

4 What days together had a total of 14 tickets sold? _____

5 If there are 42 students in the 6th grade how many are not going to the dance? _____

Directions: Use the line graph to answer questions 6 - 10.

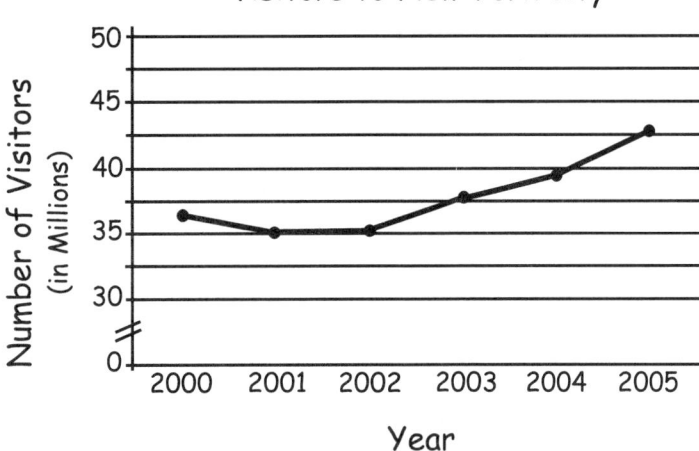

Visitors to New York City

6   Between which two years did the number of visitors decrease? _____ and _____

7   About how many visitors came to New York City in 2004? _____

8   Between which two years were the visitors about the same? _____ and _____

9   About how many more visitors came to New York City in 2005 than in 2004? _____

10   How many people do you predict will visit New York City in 2006? _____

9.4 PREDICTIONS MADE FROM DATA

To be able to determine the next step or value, based on evidence or a pattern is called a
(1) _____.

 **Hint:** You can make predictions from data that is in different forms. Some examples
X-treme   include: tables and graphs.

Example 1:

Number of Hits in a Softball Season

Year	Total Hits
1996	10
1997	14
1998	18
2000	22

If Stephanie continues playing softball, how many total hits could she possibly have in 2001?

Number of Hits in a Softball Season

Year	Total Hits
1996	10
1997	14
1998	18
2000	22

Therefore, 22 + 4 = 26.
We can predict that Stephanie would have
26 hits in the 2001 season.

Example 2: Based on the data in the graph below, how many students will be in 6th grade in 2007?

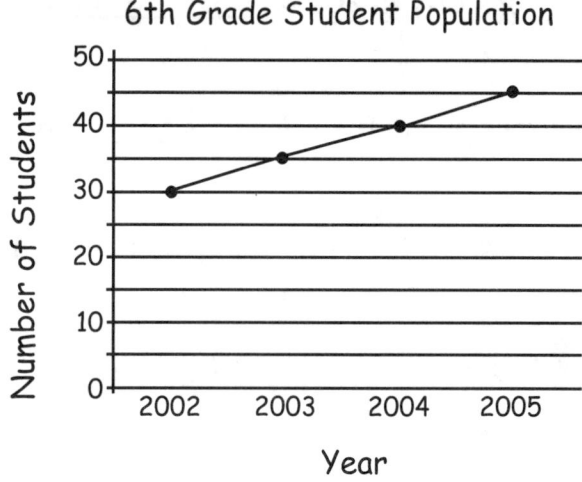

Note: Each year the number of students goes up by five students.

Based on the data in the graph there should be 55 students in 6th grade in 2007.

TEST PREP

1   Chen-Lee received a 91, 78, and 85 on three Science tests. In order to have an average of exactly 85, what grade must Chen-Lee get on the fourth test?

A   81

B   83

C   86

D   92

2   In Mrs. Renzo's class there are 18 boys and 14 girls. If Mrs. Renzo draws a name at random, what is the probability that the name will be that of a girl.

F $\frac{7}{9}$ G $\frac{9}{16}$

H $\frac{7}{16}$ J $\frac{14}{18}$

3 Nick, Andrew, Mitch, and Peter are going to a movie. Suppose the boys randomly sit in 4 seats next to each other and one of the seats is next to an aisle. What is the probability that Peter will sit in the seat next to the aisle?

A $\frac{1}{8}$

B $\frac{1}{4}$

C $\frac{1}{2}$

D $\frac{3}{4}$

4 For which set of data is there no mode?

F 2, 3, 7, 4, 0, 1, 9

G 1, 3, 2, 8, 1, 9, 7

H 6, 2, 5, 4, 7, 1, 2

J 1, 7, 9, 7, 2, 3, 4

5 For which set of data is the mean equal to the median?

A 12, 17, 23, 42

B 37, 21, 15, 45

C 18, 31, 25, 38

D 17, 24, 34, 29

Directions: Use the data from the graph to answer questions 5 and 6 in the next column..

American Pet Ownership, 2002

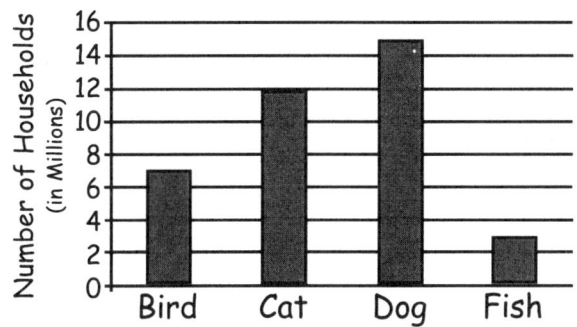

6 How many households had a cat?

F 7 million

G 12 million

H 15 million

J 3 million

7 What pet was most popular in 2002?

A Bird

B Cat

C Dog

D Fish

Directions: Use the data from the graph to answer questions 8 and 9.

Average Gasoline Prices

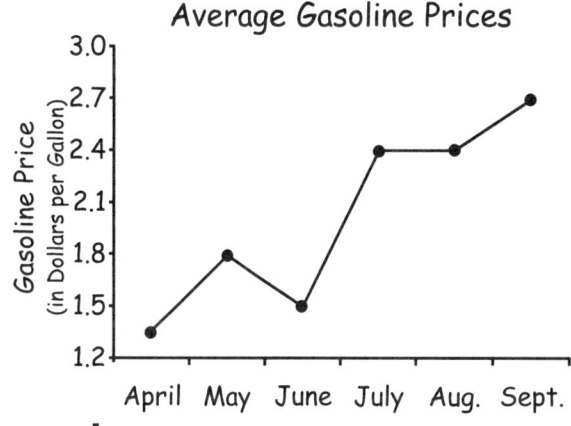

8 Which month showed a decrease in gasoline price from the previous month?

F April   G May

H June   J August

9 How much did the average gasoline price increase between June and July?

A $2.40

B $0.09

C $1.50

D $0.90

10 Several students are trying out for parts in the school musical Guys and Dolls. Kurt, David, and Joe are trying for the part of Nathan. Tina, Jun-Lee, and Laurie are trying out for the part of Adelaide.

Directions: Using a tree diagram, show how many Nathan / Adelaide pairs are possible.

Answer: _____

11 Julia was born on the 31st day of the month. Find the probability that she was born in October. **Show your work.**

Month	Jan.	Feb.	Mar.	Apr.	May.	Jun.	Jul.	Aug.	Sep.	Oct.	Nov.	Dec.
Days	31	28	31	30	31	30	31	31	30	31	30	31

Answer: _____

12 A travel agent has 10-day vacation packages to Italy for the following prices per person:
$899, $1,020, $1,350, and $1,600

Part A: Find the range of the prices. **Show your work.**

Answer: _____

Part B: Find the mean of the prices. **Show your work.**

Answer: _____

13 Reading the following data information of money collected at the telethon for a local charity, predict the amount of money that might have been raised in the year 2005. Justify your answer.

Year	Money Raised
2000	$ 152,000
2001	$ 197,000
2002	$ 242,000
2003	$ 287,000
2004	$ 332,000

Answer: _____

_____

_____

_____

_____

LESSON TEN

ESTIMATION

Vocabulary

These words and phrases are associated with Estimation and may be used when answering questions in this chapter. Definitions can be found in the Glossary/Index at the back of this *X-treme Review*.

Compatible numbers	Nonstandard measurement	Reasonable estimate
Estimate	Nonstandard unit	Round a number
Front end estimation	Personal reference	

Are We There Yet?

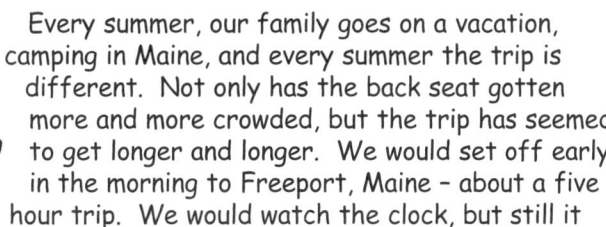

Estimation is used in your everyday life so much that you probably do not even realize you are estimating. One example of everyday estimation is traveling in the car.

Every summer, our family goes on a vacation, camping in Maine, and every summer the trip is different. Not only has the back seat gotten more and more crowded, but the trip has seemed to get longer and longer. We would set off early in the morning to Freeport, Maine – about a five hour trip. We would watch the clock, but still it always took more than 5 hours. What we didn't know – while growing up – was that it was only an estimation of how long the trip would take.

It is very easy to get Travel estimations – always available online. There are many different programs that allow you to input a start location and destination. The results are neat directions to your destination. Another feature is an approximate time that the trip will take. This approximation is based on the distance that will be traveled, and the different speed limits on each road.

Just for fun, the computer estimation from Hyde Park, New York to Freeport, Maine is 4 hours and 51 minutes (according to Yahoo Maps). In the end, I guess Mom and Dad were right it was about 5 hours.

In this lesson, you will review the different strategies used to estimate in a variety of situations. These procedures will be applied to topics that have been reviewed in previous lessons including: percent, measurement, volume, area, and circumference.

http://www.visitmaine.com
http://maps.yahoo.com

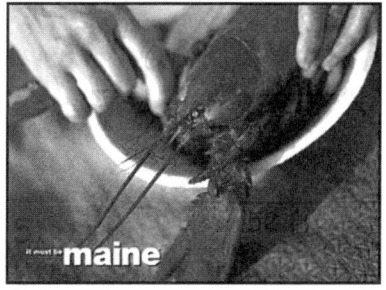

LESSON 10: ESTIMATION

10.1 ESTIMATION STRATEGIES

(1) _____ is finding the approximation of an answer. Strategies that can be used to estimate answers include: compatible numbers, front-end estimation, and rounding.

An approximation of the result of a given problem or calculation using rational, logical procedures is called a (2) _____.

(3) _____ are numbers that go together easily, a strategy often used to estimate sums, differences, products, and quotients.

Example 1: Estimate 65 ÷ 7 using compatible numbers.

$$65 \div 7$$
$$\downarrow \qquad \downarrow$$
$$64 \div 8$$

Estimated Answer: 8

1	Find numbers that are close to the ones being divided, but are easy to divide mentally.
2	Replace with the new, close numbers.
3	Divide to find the estimated quotient.

Example 2: Use compatible numbers to estimate 48.23 x 5.45

$$48.23 \quad \times \quad 5.45$$
$$\downarrow \qquad\qquad \downarrow$$
$$50 \quad \times \quad 5$$

Estimated Answer: 250

1	Find numbers that are close to the ones being multiplied, but are easy to multiply mentally.
2	Replace with the new numbers.
3	Multipiy to find the estimated product.

Hint: Compatible numbers are usually used when estimating products and quotients.

X-treme

(4) _____ is using the leading, or left-most, digits to make an estimate quickly and easily.

Example 3: Estimate 113 + 673 using front-end estimation.

$$1 + 6 = 7$$

700

700
+   80

Estimated Answer: 780

1	Add the first digit in each number.
2	Replace the first digit with the sum.
3	Add 80 because 13 + 73 is about 80.

Hint: Front-end estimation is usually used when estimating sums and differences.

To approximate the value of a whole number or decimal to a specific place value you (1) _____.

Example 4: Estimate 280 x 32 using rounding.

$$280 \quad \times \quad 32$$
$$\downarrow \qquad\qquad \downarrow$$
$$300 \quad \times \quad 30$$

Estimated Answer: 9000

| 1 | Round 280 to 300. |
| 2 | Round 32 to 30. |

Note: Round all numbers so that each contains only one non-zero digit.

PRACTICE

Directions: Use compatible numbers to estimate each product or quotient.

1   37.8 x 8.9 _____

2   63.917 ÷ 7.6 _____

3   39 ÷ 4 _____

4   77 x 21 _____

5   8.138 x 7.2 _____

6   15.83 ÷ 3.57 _____

Directions: Use front-end estimation to estimate each sum or difference.

7   387 + 925 _____

8   6003 – 5874 _____

9   6473 + 2137 _____

10   58,128 + 38,624 _____

11   8357 – 5963 _____

12   37 + 45 + 23 _____

Directions: Use rounding to estimate each sum, difference, product or quotient.

13   18 + 29 _____

14   874 ÷ 34 _____

15   78 - 24 _____

16   78 x 23 _____

17   238 ÷ 21 _____

18   98 + 38 _____

19   102 x 88 _____

20   674 - 324 _____

Directions: Explain whether or not the following estimates are reasonable. Identify which estimation strategy can be used to estimate the answer to each problem.

21 $38.8 - 15.6 \approx 25$

_____

_____

22 $\$3.87 + \$12.43 \approx \$15.00$

_____

_____

23 $21.4 \times 5.2 \approx 100$

_____

_____

24 $75.59 \div 4.23 \approx 25$

_____

_____

10.2 ESTIMATING PERCENTS

Remember: When estimating percents, use simple fractions that are close to the given values, such as halves, fourths, or tenths.

Example 1: Estimate 13 out of 45.

$\qquad$ 13 is a little more than $\frac{1}{4}$ of 45

$\qquad\qquad\qquad\qquad\qquad\qquad\qquad$ Note: $\frac{1}{4}$ = 25%

Estimated Answer: 13 out of 45 is about 25%

Example 2: Estimate what percent of the grid is shaded.

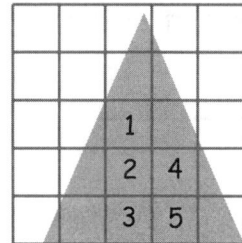

1	Count the number of squares that are completely filled in.
2	Estimate the partially filled in squares.
3	Add the estimate to the number of squares that are completely filled in.

5 squares are filled in completely.
The rest of the shading fills in about 5 more squares.

Estimated Answer: 10 out of 25 squares or $\frac{2}{5}$ which is 40%

PRACTICE

Directions: Estimate the percent.

1 2 out of 100 _____

2 32 out of 60 _____

3 71 out of 97 _____

4 280 out of 600 _____

5 30 out of 90 _____

6 18 out of 182 _____

Directions: Estimate what percent of the grid is shaded.

7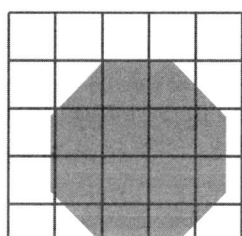

8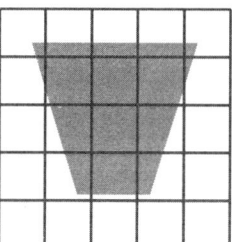

9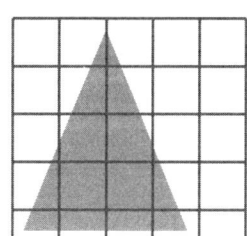

10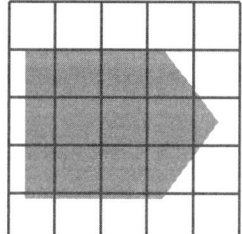

10.3 – MEASUREMENT ESTIMATION

(1) _____ is something that a person can refer to as a standard, for the purpose of comparison.

Example 1: The width of your pinky finger is about 1 cm.

The use of items as measurement tools that are not uniform in size is called
(2) _____.

Example 2: If you use fingers to measure something; one person's fingers are not necessarily the same size as another person's fingers.

Any tangible item that can be used to measure something is called a _____.

Example 3: paper clips, crayons

PRACTICE
Directions: If you did not have the proper tools, what could you use to measure the length of the following objects.

1 Length of a desk _____

2 Your height _____

3 Length of a paperclip _____

4 Height of a horse _____

10.4 ESTIMATE VOLUME, AREA, AND CIRCUMFERENCE

To estimate the volume of a rectangular prism first round the length, width, and height of the figure, and then continue with the steps from Lesson 6.

Remember: The volume of a prism is the product of the length (*l*), the width (*w*), and the *X-treme*   height (*h*)

$$V = lwh$$

Remember: The symbol for approximately, or almost equal is ≈.
X-treme

Example: Estimate the volume of the prism.

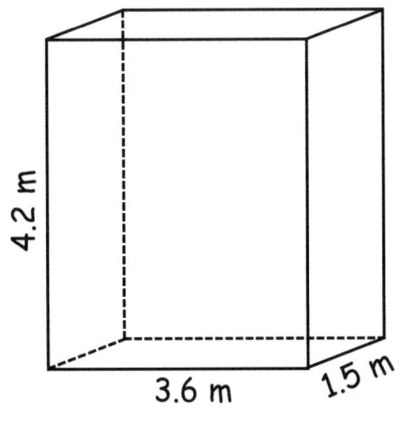

Round the length to 4.
Round the width to 2.

→

Round the height to 4.

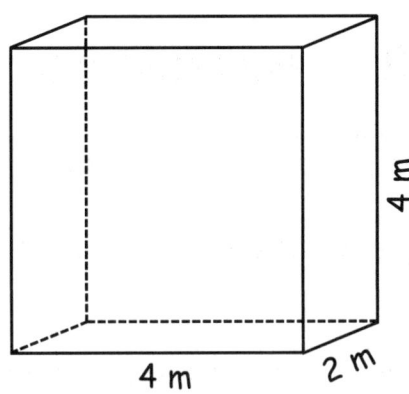

V = *lwh*	⟶	Write the formula.
V ≈ 4 · 2 · 4	⟶	Substitute.
V ≈ 8 · 4	⟶	Simplify.
V ≈ 32 m³		Cube the units.

To estimate the area, first round the dimensions of the object, and then continue with the steps from Lesson 6.

Remember:

Area of a parallelogram $A = bh$

Area of a triangle $A = \frac{1}{2}bh$

Area of a trapezoid $A = \frac{1}{2}h(b_1 + b_2)$

Example 1: Estimate the area of a triangle.

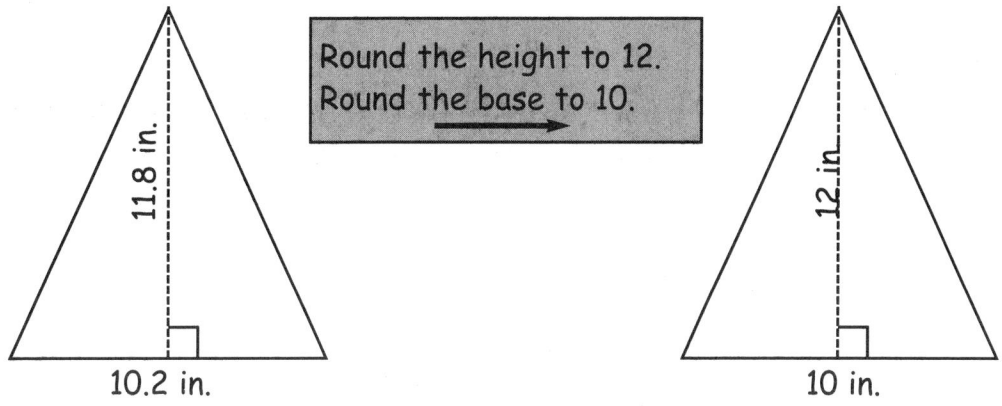

Round the height to 12.
Round the base to 10.

11.8 in.

10.2 in.

12 in.

10 in.

$A = \frac{1}{2}bh$ ⟶ Write the formula.

$A \approx \frac{1}{2} \cdot 12 \cdot 10$ ⟶ Substitute the values for the base and height.

$A \approx 6 \cdot 10$ ⟶ Simplify.

$A \approx 60 \text{ in.}^2$ ⟶ Square the units.

Example 2: Estimate the area of a square.

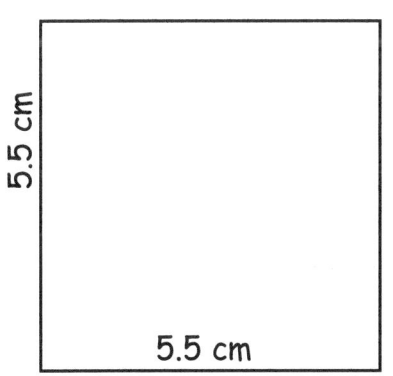
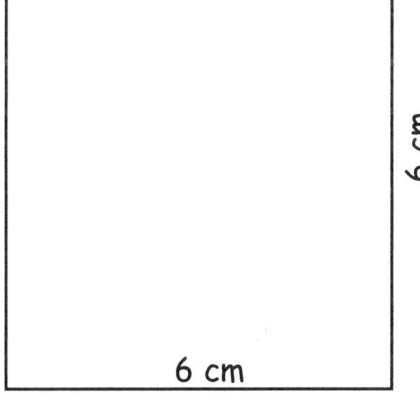

Round the sides to 6.

5.5 cm

5.5 cm

6 cm

6 cm

$A = s \cdot s$ ⟶ Write the formula.

$A \approx 6 \cdot 6$ ⟶ Substitute the values for the sides.

$A \approx 36 \text{ cm}^2$ ⟶ Simplify and Square the units.

Example 3: Estimate the area of a parallelogram.

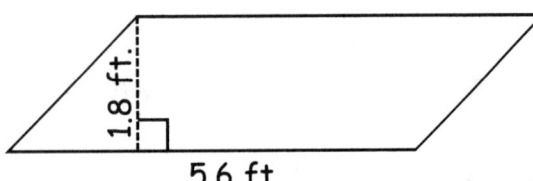   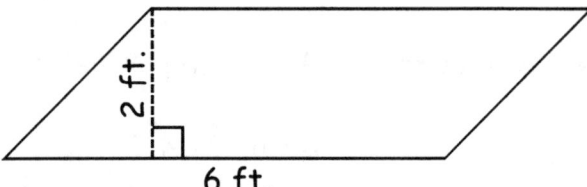

Round the height to 2.
Round the base to 6.
⟶

$A = bh$ ⟶ Write the formula.
$A \approx 6 \cdot 2$ ⟶ Substitute the values for the base and height.
$A \approx 12$ ft^2 ⟶ Simplify and Square the units.

Example 4: Estimate the area of a trapezoid.

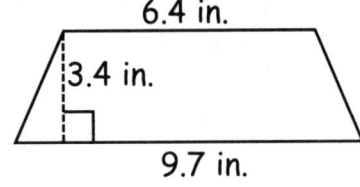   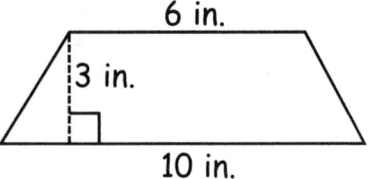

Round the height to 3.
Round the 1st base to 6.
Round the 2nd base to 10.
⟶

$A = \frac{1}{2}h(b_1 + b_2)$ ⟶ Write the formula.

$A \approx \frac{1}{2} \cdot 3(6 + 10)$ ⟶ Substitute the values for the bases, and the height.

$A \approx \frac{1}{2} \cdot 3\,(16)$ ⟶ Simplify.

$A \approx \frac{1}{2} \cdot 48$ ⟶ Simplify.

$A \approx 24$ in.^{2} ⟶ Square the units.

To estimate the circumference, first round the diameter or radius, and then continue with the steps from Lesson 7.

Remember: The ratio of every circle's circumference to its diameter is the same – π. *Xtreme* You should use 3.14159 as an approximation for π.

$$C = \pi d \qquad\qquad C = 2\pi r$$

Example 5: Estimate the circumference of the circle.

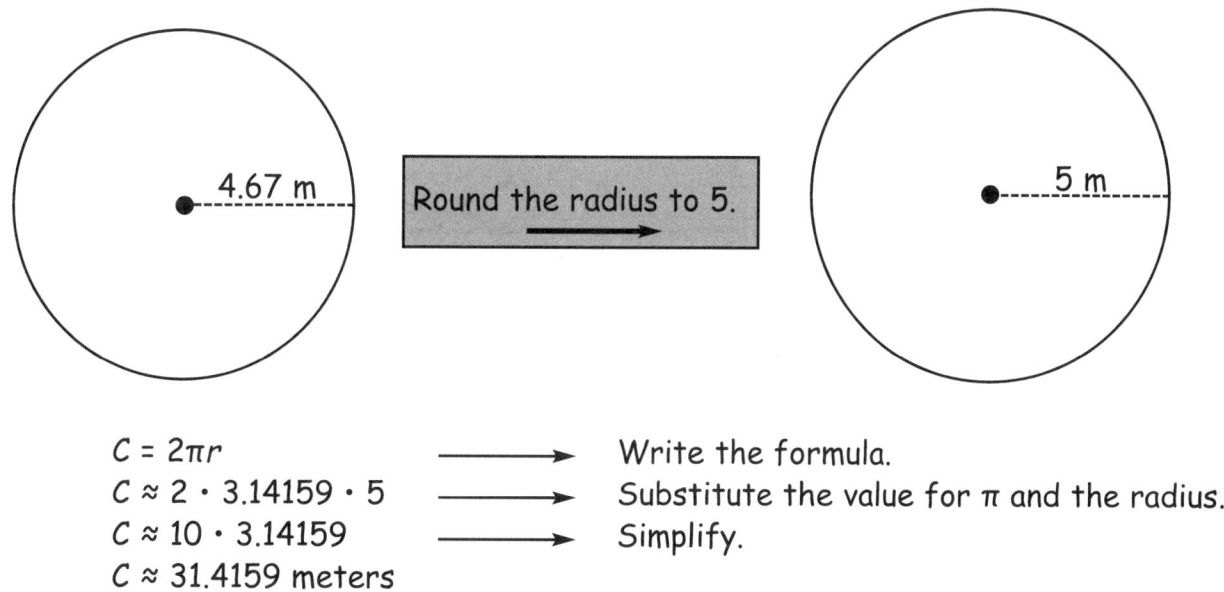

Round the radius to 5.

$C = 2\pi r$ → Write the formula.

$C \approx 2 \cdot 3.14159 \cdot 5$ → Substitute the value for π and the radius.

$C \approx 10 \cdot 3.14159$ → Simplify.

$C \approx 31.4159$ meters

PRACTICE

Directions: Estimate the volume of the following rectangular prisms [not drawn to scale].

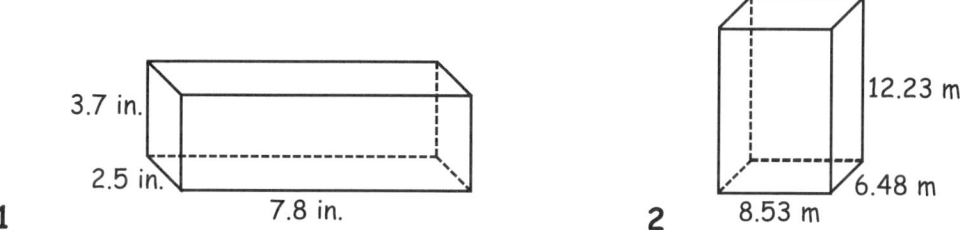

1

2

Directions: Estimate the area of the following polygons [not drawn to scale].

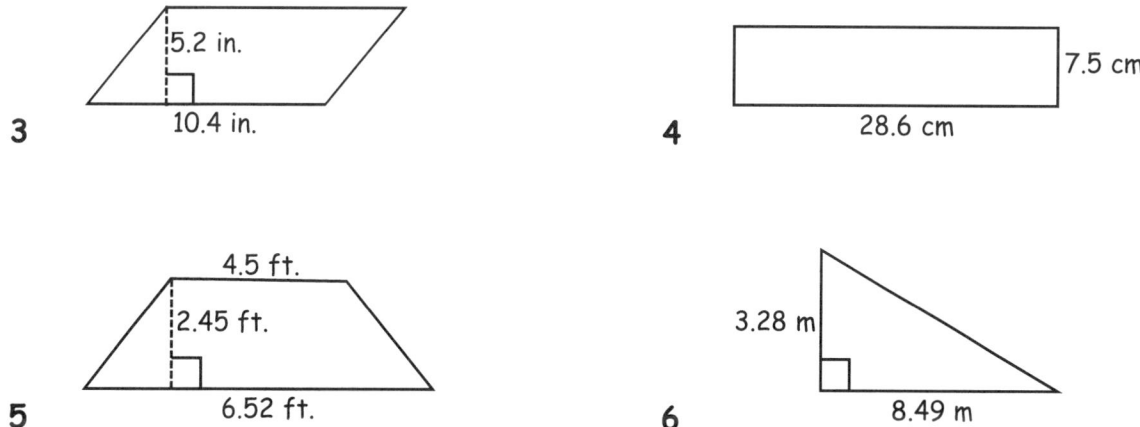

3

4

5

6

Directions: Estimate the circumference of the following circles [not drawn to scale].

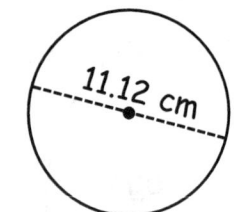

7

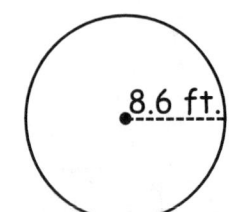

8

TEST PREP

1 Estimate what percent of the grid is shaded.

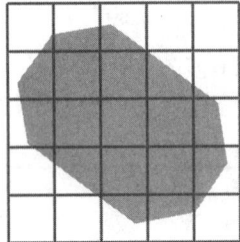

A   25%
B   48%
C   60%
D   80%

2 Find an approximation of the area of the triangle below [not drawn to scale].

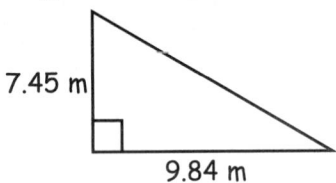

F   31.5 m²
G   35 m²
H   36 m²
J   36.654 m²

3 Estimate the difference of the following numbers:

5,493 – 1,375

A   4,000
B   4,118
C   4,500
D   5,000

4 Estimate the ciecumference of the circle below [not drawn to scale].

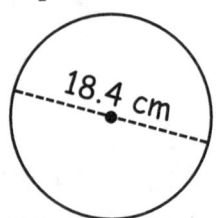

F   28 cm
G   30 cm
H   54 cm
J   70 cm

5 Estimate the sum of the following numbers:

58 + 63 + 56 + 61

A   200
B   220
C   240
D   260

6 Find the approximate volume of the rectangular prism [not drawn to scale].

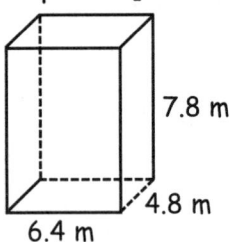

F   168 m³
G   210 m³
H   239 m³
J   240 m³

7 Al wants to buy a $17.89 CD and a $39.97 X-Box Game. Estimate the amount of money he will need.

    A $50.00

    B $56.00

    C $57.86

    D $58.00

8 Eliza rode her bike 3,260 miles across the United States in 42 days. Estimate the distance she rode each day.

    F   70 miles

    G   75 miles

    H   77 miles

    J   82 miles

9 Grace and Tess are building a cage for their new bunnies. They have all of the tools they need to build the cage except a measuring tape. What could they use to measure the building materials? Explain your answer.

_____

_____

_____

_____

_____

10 Marco has 4 dogs which are all about the same size. If the dogs weigh a total of 128.4 lbs., estimate the weight of each dog. **Show your work.**

Answer: _____

11 Amanda went shopping for a new outfit for school with $65. She found a shirt for $26.99, and pants for $32.45 (taxes included). Just before she checked out she estimated the cost of her outfit.

Amanda thinks she has enough money because she estimated the cost of her outfit to be $60. How did Amanda estimate the cost of her outfit, and was her estimate correct?

_____

_____

_____

_____

12 Peter is putting in grass on the side of his property. He needs to make sure he has enough grass seed to layout in the entire area.

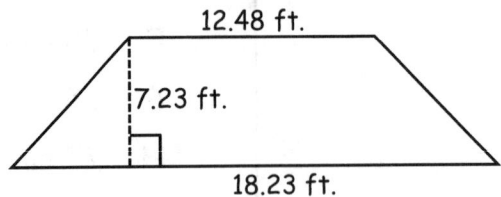

12.48 ft.

7.23 ft.

18.23 ft.

Part A: Estimate the area of the lawn. [not drawn to scale] **Show your work.**

Answer: _____

Part B: Explain why you would want to estimate the area.

_____

_____

_____

_____

11
X-treme

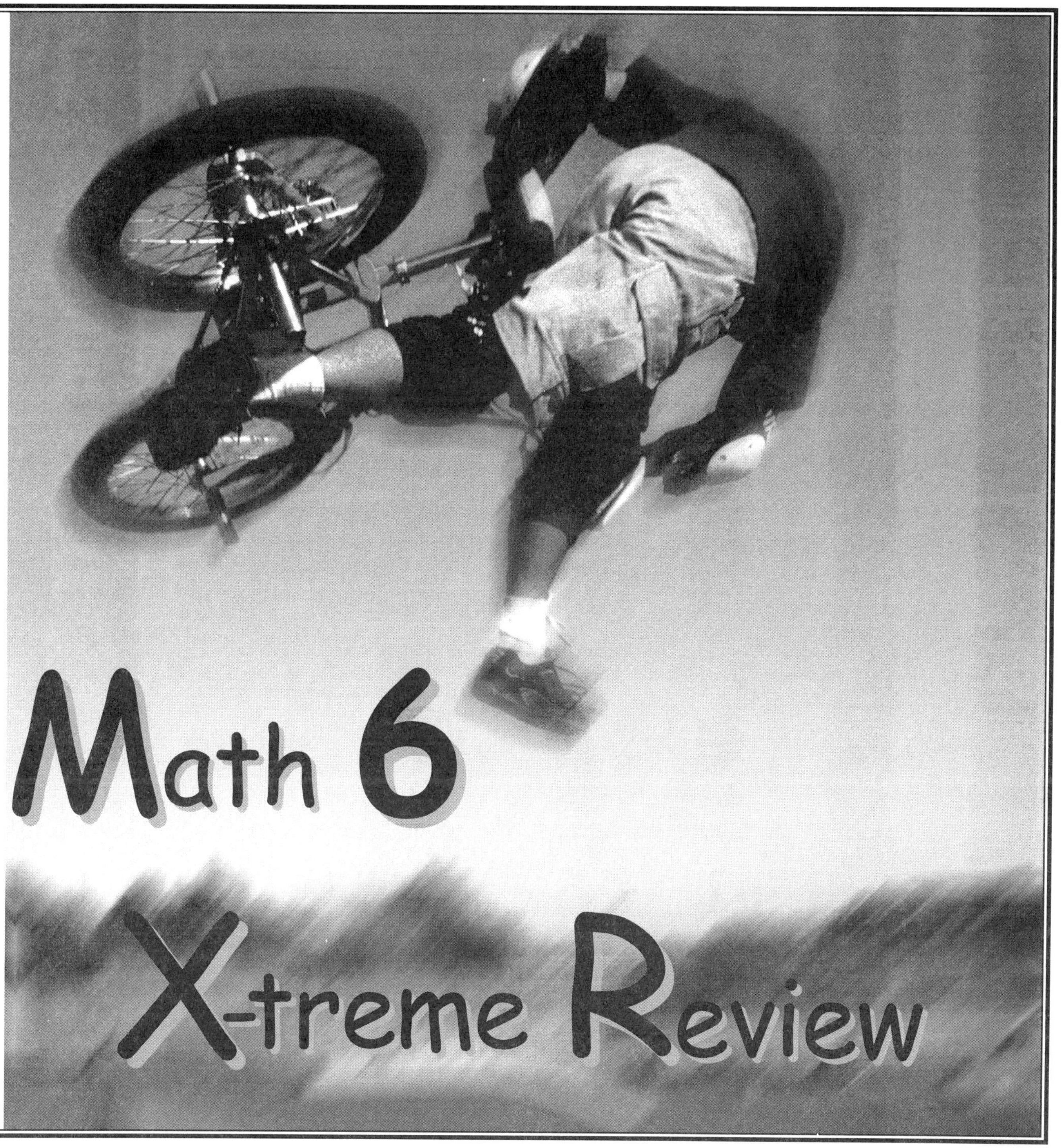

Math 6

X-treme Review

MATH 6 – *X-TREME REVIEW*

Book 1

1 At Midtown Middle School, 14 out of 35 students in the marching band are in 6th grade. Which other ratio is equivalent to the number of 6th graders in the marching band.

A $\frac{2}{5}$

B $\frac{5}{7}$

C $\frac{21}{35}$

D $\frac{28}{35}$

2 Bobby drew two rectangles and one square on the grid below.

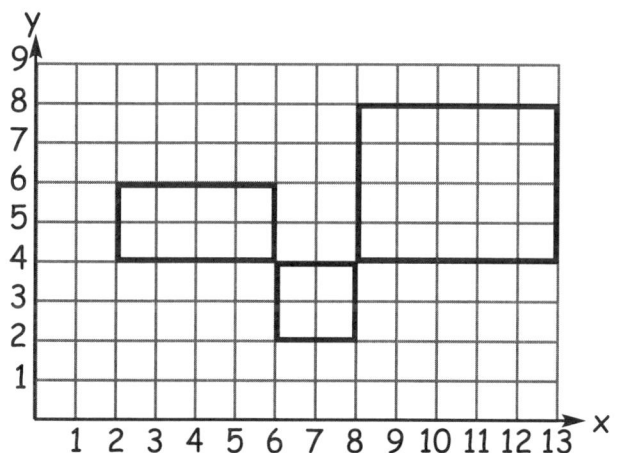

What is the total perimeter, in units, of the three shapes?

F 16

G 22

H 32

J 38

3 Ricky jogs $\frac{1}{2}$ of a mile. What is another way to write this number?

A 0.15

B 0.2

C 0.3

D 0.5

4 Simplify the expression below.

$$7^2 - 3^3$$

F 4

G 5

H 12

J 22

5 What line segment represents a radius of the circle below?

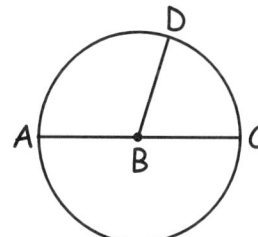

A $\overline{AD}$

B $\overline{BC}$

C $\overline{DC}$

D $\overline{AC}$

6 The expression for Kaitlyn's age is $m - 11$, where m is Max's age. How old is Kaitlyn if Max is 24?

F 11

G 13

H 24

J 35

7   Emily bought 36 eggs. Seven of them were brown. Five of them were spotted. The rest were white. If she chooses an egg at random, what is the probability that she will choose a white egg?

A $\frac{7}{36}$

B $\frac{5}{18}$

C $\frac{2}{3}$

D $\frac{1}{2}$

8   Ann is building an ice sculpture out of ice cubes.

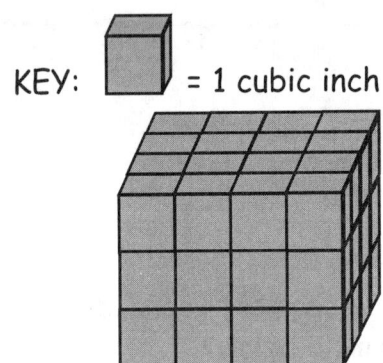

KEY: = 1 cubic inch

What is the volume of Ann's ice sculpture?

F   10 cubic inches
G   12 cubic inches
H   30 cubic inches
J   48 cubic inches

9   Scott bought 48 pints of ice cream for a birthday party. How many gallons of ice cream did Scott buy?

1 gallon = 4 quarts
1 quart = 2 pints

A   3
B   6
C   12
D   24

10 A recipe for pie crust calls for $\frac{2}{3}$ cup of butter. How much butter would you use to make $\frac{1}{2}$ of the crust in the original recipe.

F $\frac{1}{2}$

G $\frac{3}{5}$

H $\frac{2}{6}$

J $\frac{4}{6}$

11 Which expression would you use to find the number of cows in a barn if you counted h hooves?

A $h + 4$
B $h - 4$
C $4h$
D $\frac{h}{4}$

12 The distance down the length of a football field including the end zones is 360 feet. What is the length of the field in yards?

1 yard = 3 feet

F   90
G   120
H   180
J   1080

13 Sam had 76 baseball cards. His little sister Hannah destroyed 25% of them. How many baseball cards did Hannah destroy?

A   19
B   25
C   51
D   56

14 Mary painted $\frac{1}{5}$ of the wall of her TV room light blue. Then, She painted $\frac{2}{6}$ of another section of the same wall light yellow. What fraction of the wall is not painted?

F $\frac{7}{15}$

G $\frac{10}{30}$

H $\frac{16}{30}$

J $\frac{8}{15}$

15 Linda is drawing a trapezoid on the grid below by plotting a fourth point and then connecting all of the points.

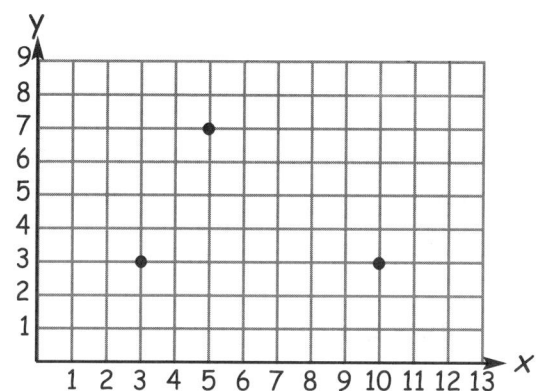

Which coordinates would best complete the trapezoid?

A (5, 7)
B (6, 8)
C (8, 7)
D (8, 6)

16 Which equation shows the commutative property of multiplication?

F 7 x 2 = 7 x 2
G 7 x 2 = 14 ÷ 2
H 7 x 2 = 3 x 4
J 7 x 2 = 2 x 7

17 Carson is putting in a walkway made with triangular stones, like the one drawn below. [not drawn to scale]

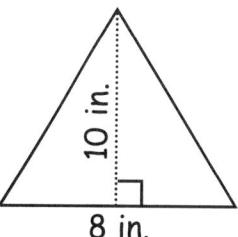

10 in.

8 in.

What is the area, in square inches, of the triangular stone?

A 18
B 40
C 60
D 80

18 Eileen cut five pieces of ribbon measuring:

$1\frac{1}{4}$ ft., $\frac{1}{2}$ ft., $2\frac{3}{4}$ ft., $\frac{2}{3}$ ft., $1\frac{1}{3}$ ft.

Which list shows the lengths of the five ribbons in order from shortest to longest?

F $\frac{1}{2}$ ft., $\frac{2}{3}$ ft., $1\frac{1}{3}$ ft., $1\frac{1}{4}$ ft., $2\frac{3}{4}$ ft.

G $2\frac{3}{4}$ ft., $1\frac{1}{3}$ ft., $1\frac{1}{4}$ ft., $\frac{2}{3}$ ft., $\frac{1}{2}$ ft.

H $\frac{2}{3}$ ft., $\frac{1}{2}$ ft., $1\frac{1}{3}$ ft., $1\frac{1}{4}$ ft., $2\frac{3}{4}$ ft.

J $\frac{1}{2}$ ft., $\frac{2}{3}$ ft., $1\frac{1}{4}$ ft., $1\frac{1}{3}$ ft., $2\frac{3}{4}$ ft.

19 Veronica writes the expression: 4 x 4 x 4 x 4 x 4. Which is another way to write the same expression using exponents?

A 4^4
B 4^5
C 5^4
D 5^5

20 The graph below shows the number of pets owned by Mrs. Hill's students.

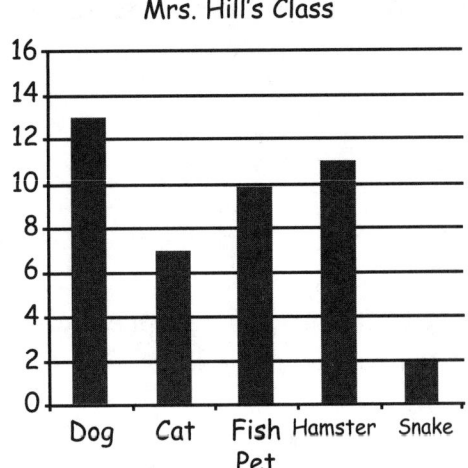

Mrs. Hill's Class

What is the total number of pets owned by Mrs. Hill's students?

F 13
G 20
H 31
J 43

21 Adrian asked seven of his classmates how many movies they saw last year. The number of movies each classmate saw is recorded below.

12, 2, 10, 4, 8, 7, 10

What is the mode number of movies seen?

A 7
B 8
C 10
D 12

22 Cara's solution for an equation is 15. Which equation could she have solved?

F $5 + x = 75$
G $5 - x = 75$
H $5x = 75$
J $\dfrac{2}{x} = 75$

23 Hope recorded the different flowers in her garden. She organized the results in the table below.

FLOWERS IN THE GARDEN

Flower	Number of Flowers
Daisy	12
Carnation	6
Marigold	16
Rose	4

What fraction of the flowers in Hope's garden are marigolds?

A $\dfrac{2}{19}$

B $\dfrac{3}{19}$

C $\dfrac{6}{19}$

D $\dfrac{8}{19}$

24 A 30 ounce box of cornflakes contains 20 servings. Each serving is $1\frac{1}{2}$ ounces. If seven servings are used how much cereal is left in the box?

F $19\frac{1}{2}$ ounces

G $17\frac{3}{4}$ ounces

H 13 ounces

J 23 ounces

25 What is the value of the expression below when $n = 12$?

$$8 - 2 \cdot \dfrac{n + 2}{7}$$

A 2
B 4
C 8
D 12

Book 2

26 Paul's sock drawer contains 29 pairs of socks. There are 8 black pairs and 21 white pairs of socks.

Part A
Estimate the percent of white pairs in the drawer.

Show your work.

Estimate: _____ %

On the lines below, use words, symbols, or numbers to explain how to estimate the percent of white socks in the drawer.

_____

_____

_____

_____

Part B
Paul adds 12 blue socks to the drawer. Estimate the percent of socks in the drawer that are blue.

Show your work.

Estimate: _____ %

27 New York City received 2 inches of rain on Monday, 5 inches on Tuesday, 1 inch on Wednesday, no rain on Thursday and Friday, $1\frac{1}{2}$ inch on Saturday, and $2\frac{1}{2}$ inches on Sunday.

Part A

What percent of the rainfall for the week fell on Monday?

Show your work.

Answer: _____ %

Part B

What percent of the total rainfall for the week, fell Monday through Wednesday?

Show your work.

Answer: _____ %

28 Catherine wants to determine the area of her bedroom. A diagram of the bedroom is shown below.

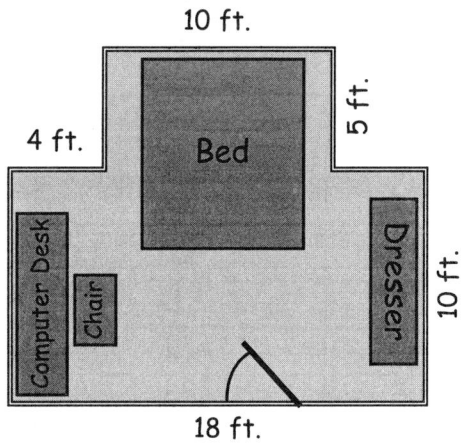

[not drawn to scale]

What is the area, in square feet, of the bedroom?

Show your work.

Answer: _____ square feet

On the lines below, explain how you determined the area.

_____

_____

_____

_____

29 George is getting ready to try out for the school swim team. He is swimming laps daily and increases his number of laps each day by following a number pattern. The number of laps George does for 7 days is shown in the table below.

GEORGE'S LAPS

Day	Number of Laps
1	3
2	7
3	11
4	15
5	19
6	23
7	27

Part A

If George continues to swim laps according to the number pattern, how many laps will he do on day 12?

Show your work.

Answer: _____ laps

Part B

George's tryout is on day 23. He needs to be able to swim 80 laps. On the lines below, use words, symbols, or numbers to explain whether George will be ready for the tryout.

_____

_____

_____

_____

30 Jake has $65 saved in the bank. He plans on adding $5 each week to his savings.

Write an expression for the amount of money Jake will have in his bank account at any time.

Expression: _____

On the lines below, explain how you determined your expression.

_____

_____

_____

_____

31 Brian's age is 4 years older than Diane who is *n* years old.

Write an expression to represent Brian's age.

Expression: _____

If Diane is 10 years old, how old is Brian?

Show your work.

Answer: _____

32 Jenna took a survey of her classmates. She asked each classmate the ages of their siblings. Her results are shown in the tally chart below.

SIBLINGS AGES

Age	Frequency
1 - 4	2
5 - 8	7
9 - 12	3
13 - 16	5
17 - 20	3
21 - 24	1

Part A

What fraction of the siblings are in the age group 1-4? Write your fraction in lowest terms.

Show your work.

Answer: 1-4: _____

Part B

What fraction of the siblings are in the age groups 5-8 and 13-16? Write your fractions in lowest terms.

Show your work.

Answer: 5-8: _____

Answer: 13-16: _____

33 Michaela is getting some new fish. Her grandfather gave her a fish tank for her birthday, like the one shown below. [not drawn to scale]

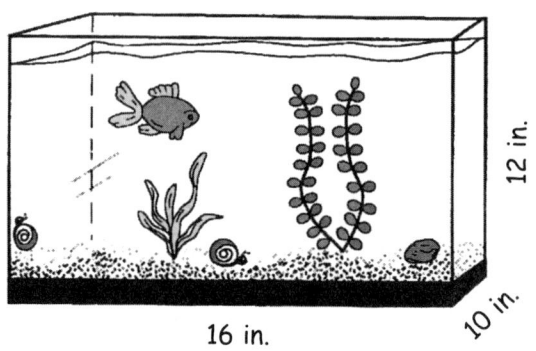

16 in.

12 in.

10 in.

What is the volume, in cubic inches, of the fish tank?

$$V = lwh$$

Show your work.

Answer: _____ cubic inches

34 Simplify the expression below.

$$3 \times 8 \div 4 + 23$$

Show your work.

Answer: _____

35 Mateo is painting a 14 inch diameter wheel for one of the games at the school carnival, as shown below. [not drawn to scale]

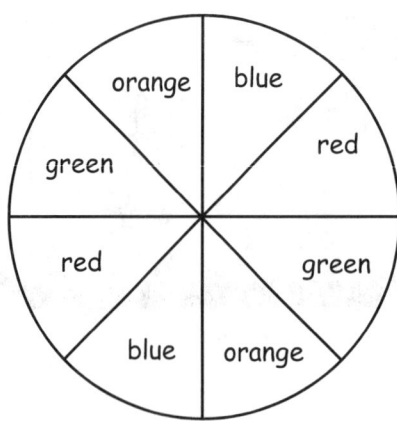

Mateo needs to find the area of the orange sections of the wheel in order to buy the right amount of paint. What is the area of the orange sections of the wheel? Use 3.14159 as the value of π. Give your answer in square inches and round your answer to the nearest hundredths place.

$$A = \pi r^2$$

Show your work.

Answer: _____ in.^{2}

Session 1

1 What number is represented by point Z on the number line below?

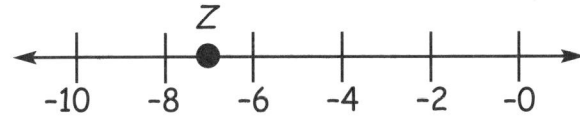

- **A** -6
- **B** -7
- **C** -8
- **D** -9

2 Carly divides s strips of film into 8 sections. She then separates all of the sections into 4 piles. Which expression represents the number of sections in each of the 4 piles created?
- **F** $s \div 8 \div 4$
- **G** $8s \div 4$
- **H** $(s \div 8) \div 4$
- **J** $(4 \div 8)s$

3 Which equation is true when $n = 2$?
- **A** $8 + n - 4 = 5$
- **B** $2 + 6 - n = 4$
- **C** $n - 2 + 5 = 8$
- **D** $7 + n - 3 = 6$

4 Chloe's new swimming pool has a diameter of 16 feet. What is the radius of the swimming pool?
- **F** 2 feet
- **G** 4 feet
- **H** 8 feet
- **J** 32 feet

5 Duane rolled a six sided die 15 times. He recorded each roll of the die. The results are shown in the table below.

ROLLS OF THE DIE

Side	Number of Times
1	I I
2	
3	I I I
4	I I I I
5	I
6	╫╫

What fraction of the die rolls were even numbers?

- **A** 0
- **B** $\dfrac{4}{15}$
- **C** $\dfrac{2}{15}$
- **D** $\dfrac{3}{5}$

6 Two similar rectangles are shown below.

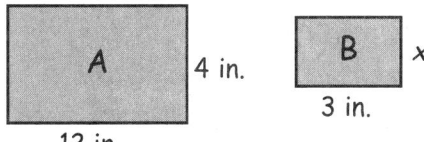

What is the length of side x in rectangle B?

- **F** $\dfrac{1}{2}$
- **G** 1
- **H** 4
- **J** 8

7 Simplify the expression below.

$$4^3 - 7^2$$

A -2
B 3
C 5
D 15

8 What is the volume of the package below?

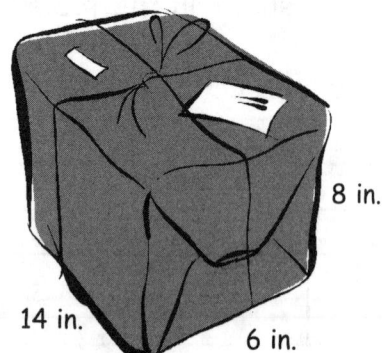

8 in.

14 in.

6 in.

F 28 cubic inches
G 336 cubic inches
H 672 cubic inches
J 1,344 cubic inches

9 What value for *h* makes the equation true?

$$2h + 3 = 15$$

A 3
B 6
C 9
D 12

10 Greta is painting different shapes on the playroom wall. She has outlined a triangle that had a height of 5 feet, and a base that measured 6 feet. What is the area of the triangle that Greta will paint?

$$A = \frac{1}{2}bh$$

F $5\frac{1}{2}$ square feet
G 11 square feet
H 15 square feet
J 30 square feet

11 Simplify the expression below.

$$(9 + 2)^2 - 17$$

A 5
B 36
C 66
D 104

12 Jenny is inviting 35 people to her birthday party. She has written out 14 of the invitations so far. What percent of the total number of invitations does Jenny still have to write out?

F 21%
G 35%
H 40%
J 60%

13 Abigail recorded the number of tickets sold each night at the local movie theater seven nights. The results are shown below.

42, 60, 36, 52, 48, 36, 42

What is the median number of tickets sold at the movie theater?

A 24
B 36
C 42
D 53

14 Which point on the number line is less than -2 but greater than -8?

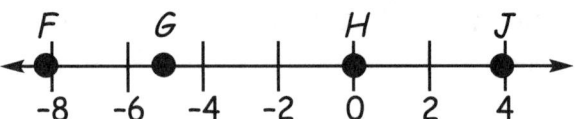

F F
G G
H H
J J

15 The gas tank on Dan's lawn mower holds 3 pints of gasoline. The refill container holds 4 gallons of gasoline. How many times can Dan refill the tank on the lawn mower with the gasoline in the refill container?

> 1 gallon = 4 quarts
> 1 quart = 2 pints

A   5
B   7
C   10
D   16

16 Mandy wrote the expression below.

$$\frac{7g + 4}{5}$$

If g equals 8, what is the value of the expression?
F   4
G   10
H   12
J   15

17 Which equation shows the identity property (element) of addition?
A $7 + 1 = 8$
B $7 + 1 = 1 + 7$
C $7 + 0 = 0 + 7$
D $7 + 0 = 7$

18 At a skateboard competition, $\frac{4}{10}$ of the events have taken place. What percent of the competition is left.
F   20%
G   40%
H   60%
J   80%

19 Jan needed 6 quarts of lemon juice to make lemonade for the neighborhood barbeque. How many cups of lemon juice will she need?

> 1 quart = 2 pints
> 1 pint = 2 cups

A   6
B   12
C   24
D   36

20 A circle has a radius, $\overline{ST}$, as shown below.

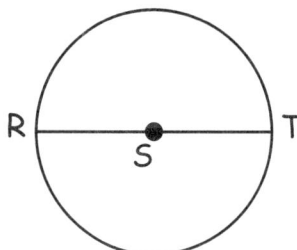

The diameter $\overline{RT}$ is 15 inches. What is the length of $\overline{RS}$?

F   3 inches
G   6 inches
H   7.5 inches
J   15 inches

21 Of Ray's plants, $\frac{1}{3}$ are tulips and $\frac{2}{5}$ are daisies. What fraction of Ray's garden are neither tulips nor daisies?

A $\frac{2}{3}$

B $\frac{2}{15}$

C $\frac{4}{15}$

D $\frac{11}{15}$

22 Janet has 6 red cards, 10 yellow cards, and 8 green cards. If she chooses one card at random what is the probability that she will choose a green card?

F $\frac{1}{4}$

G $\frac{2}{15}$

H $\frac{4}{15}$

J $\frac{1}{3}$

23 Nicole is bringing 2,500 milliliters of soda to a party. How many liters of soda is Nicole bringing to the party?

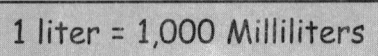
1 liter = 1,000 Milliliters

A 2.5
B 5.0
C 50
D 250

24 Bart plotted four points in Quadrant I of the grid below.

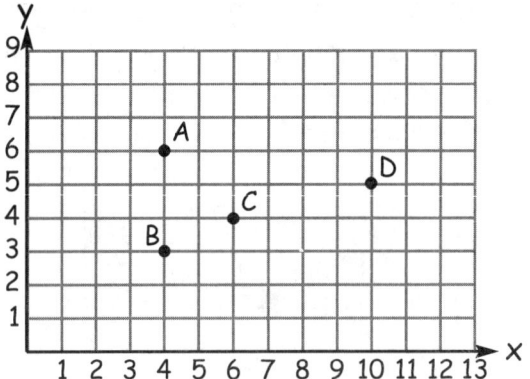

What coordinates for point A?
F (4, 3)
G (4, 6)
H (6, 4)
J (10, 5)

25 Which equation can be used to help solve the equation below?

$$c \div 12 = 728$$

A c = 728 + 12
B c = 728 - 12
C c = 728 x 12
D c = 728 ÷ 12

Session 2

26 The Zambroski family is having a garage sale. Sally visits the garage sale and wants to buy a lamp for $16.25, a bookcase for $24.50 and a mirror for $7.25.

Part A
By rounding to the nearest dollar, **estimate** the amount of money Sally will need to purchase the lamp, bookcase, and mirror. **Show your work.**

Answer: $_____

Part B

Sally spots a vase that costs $1.95. If she has $50, can she afford to buy the vase too?

On the lines below, explain your answer.

_____

_____

_____

_____

27 The letters of the word "homework" are written on a set of cards:

H O M E W O R K

The cards are shuffled and placed face down on a table.

Part A

If Jamie randomly picks a card, what is the probability that the card will be a <u>vowel</u>?

Probability: _____

Part B

If Jamie randomly picks a card, what is the probability that the card she picks will be an *O* ?

Probability: _____

On the lines below, explain your answer.

_____

_____

_____

_____

28 There are 20 buttons in a box. Twenty percent (20%) of the buttons are made of brass. Besides the brass buttons, the box includes 6 pearl buttons, 3 cloth buttons, 3 glass buttons, and some plastic ones.

Part A

How many buttons are made of brass?

Show your work.

Answer: _____ brass buttons

Part B

If the brass buttons are removed from the box, then $\frac{1}{4}$ of the remaining buttons are made of plastic. How many buttons are made of plastic?

Show your work.

Answer: _____ plastic buttons

29 Order the following expressions from least to greatest:

$$3^3, 5^2, 2^3, 7^2, 3^2, 2^2$$

Show your work.

Answer: _____

NO PERMISSION HAS BEEN GRANTED BY N&N PUBLISHING COMPANY, INC TO REPRODUCE ANY PART OF THIS BOOK BY ANY MECHANICAL, PHOTOGRAPHIC, OR ELECTRONIC PROCESS.

PAGE 142   MATH 6 – *X-TREME REVIEW*   N&N©

30 Thomas planted a 4 foot tall Maple tree in his backyard. The garden center predicts that the tree will grow 2 feet each year.

Part A

Write an expression for the height after *y* years.

Expression: _____

Part B

Solve your expression using 5 years as the value of *y*.
Show your work.

Answer: _____

31 Solve the equation below for *k*.

$$k + 7 = 54$$

Show your work.

Answer: *k* = _____

32 Joel wants to buy more geckos for his terrarium, shown below. [not drawn to scale]

30 cm

20 cm

65 cm

Part A

What is the volume of the terrarium? Use the formula V = *lwh*.

Show your work.

Answer: _____ cubic centimeters (cm³)

Part B

A gecko about the same size as the one he has now, needs at least 14,000 cm³ of space. How many more geckos can fit in Joel's terrarium? **Show your work.**

Answer: _____

On the lines below, explain how you found your answer.

_____

_____

_____

_____

33 Use the grid below.

Part A
On the grid, draw two different rectangles, each with a perimeter of 20 units.

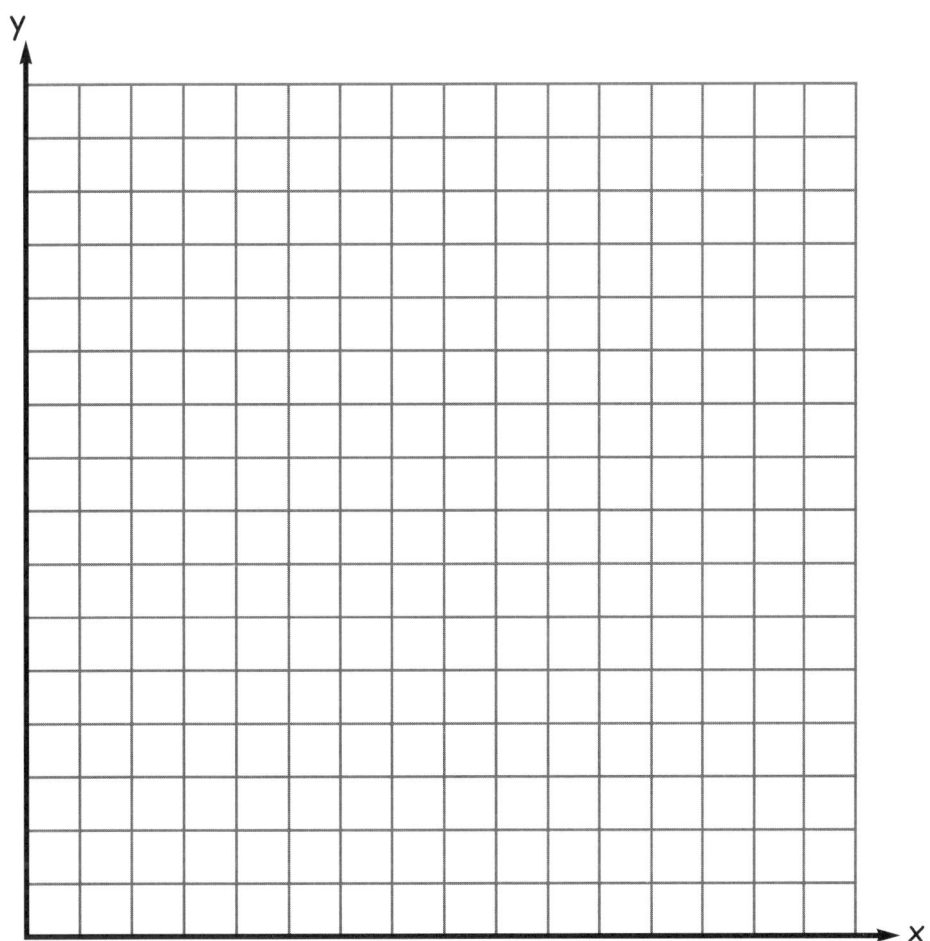

Part B
What are the lengths of each side of the rectangles?

Answer: Rectangle 1: _____ units

Rectangle 2: _____ units

34 Of the top 40 songs of the month, 24 of them were recorded by male artists. What percent of the top 40 is this?

Show your work.

Answer: _____ %

35 A diagram of Alice's garden is shown below. [not drawn to scale]

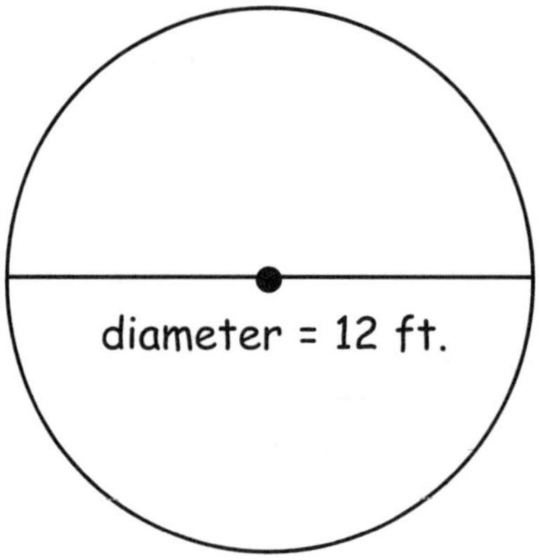

diameter = 12 ft.

Alice wants to put up a fence around her garden to keep out the rabbits. What is the circumference of Alice's garden?
Show your work.

Answer: _____ feet

On the lines below, explain your answer.

_____

_____

_____

_____

This reference section is a combination word definition Glossary and "look up" page Index.

The Index portion of this section identifies which Lesson and on what page a particular glossary word and/or subject is found.

The Vocabulary Definitions found in this Glossary and Index can help you with understanding the Lesson Vocabulary and in answering questions in the Lessons of this *Math 6 X-treme Review*.

Absolute value [pg 14] A non-negative number equal in numerical value to a given real number.

addition [pg 9] The operation of combining numbers so as to obtain an equivalent total.

additive inverse [pg 9] A number that when added to a given number results in a sum of zero; the opposite of a number.

algebra [pg 5] The branch of mathematics that uses letters, symbols, and/or characters to represent numbers and express mathematical relationships.

algebraic [pg 50] Making use of or referring to concepts or methods of algebra.

algebraic expression [pg 50-52] A mathematical phrase that is written using one or more numbers, variables, constants, signs of operation, and symbols of algebra. (e.g., $3y + 6$).

algebraic solution [pg 49] The process of solving a mathematical problem using the principles of algebra.

approximation [pg 112] A mathematical quantity that estimates a desired quantity.

arc [pg 86] Part of a curve between any two of its points (e.g., the arc of a circle or sector).

area [pg 70, 84, 118] The measure of the interior surface of a closed region or figure; area is measured in square units (e.g., ft^2, cm^2).

associative property [pg 10] A property of real numbers that states that the sum or product of a set of numbers is the same, regardless of how the numbers are grouped.

axis [pg 60, 63] A horizontal or vertical line used in the coordinate system used to locate a point on the coordinate graph.

axes [pg 60, 63] The horizontal and vertical lines dividing a coordinate plane into four quadrants.

Base [pg 30, 42] A number that is multiplied by a rate or of which a percentage or fraction is calculated.

base-ten number system [pg 92] A place value number system in which ten digits, 0 through 9, are used to represent a number and the value of each place is 10 times the value of the place to its right; the value of any digit in the number is the product of that digit and its place value.

bases [pg 95] The top and bottom of a rectangular prism (symbol: b_1, b_2).

Calculate [pg 29] To compute; to perform the indicated operation(s).

capacity [pg 92-95] The maximum amount a container can hold.

coordinate plane [pg 60, 63] The plane formed by a horizontal axis and a vertical axis, often labeled the *x*-axis and *y*-axis respectively.

central angle [pg 86] An angle whose vertex is at the center of a circle and whose sides contain radii of the circle.

chord [pg 82, 85] A line segment connecting any two points on a circle.

circle [pg 82] A plane closed curve consisting of all points a fixed distance from a fixed point called its center.

circumference [pg 84, 118] The distance around a circle, calculated by multiplying the length of the diameter of the circle by pi (i.e., $C = \pi d$).

common denominator [pg 12-13] A common multiple of the denominators of a number of fractions.

commutative property of addition [pg 10] A property of real numbers that states that the sum of two terms is unaffected by the order in which the terms are added; i.e., the sum remains the same (e.g., $a + b = C$; $b + a = C$).

commutative property of multiplication [pg 10] A property of real numbers that states that the product of two factors is unaffected by the order in which they are multiplied; i.e., the product remains the same (e.g., $3 \times 5 = 5 \times 3$ and $5 \cdot x = x \cdot 5$).

compatible numbers [pg 112] Numbers that go together easily, used to estimate sums, differences, products, and quotients.

congruent figures [pg 22, 25, 73] Polygons having the same size and shape (symbol: $\cong$).

coordinate axes [pg 60] The two intersecting perpendicular lines in a plane that form the four quadrants for locating points, given the ordered pair of the points; the axes are referred to as the x-axis and the y-axis.

coordinate geometry [pg 60] The study of geometry using a coordinate plane.

coordinates [pg 60] An ordered pair of numbers that identifies a point on a coordinate plane, written as (x, y).

corresponding angles and sides [pg 22-23] Sides in the same relative position on two congruent or similar figures. (The corresponding sides of congruent figures are equal and the corresponding sides of similar figures are proportional.)

counting numbers [pg 7, 8] All whole numbers greater than zero; also called natural numbers.

cross multiplication [pg 20, 30-31] To clear an equation of fractions when each side consists of a fraction with a single denominator by multiplying the numerator of each side by the denominator of the other side and equating the two products obtained.

cube [pg 76] A solid rectangular figure (prism) with 6 square faces, all equal in area.

customary measurement system [pg 92] The system of measurement used mainly in the United States to measure length (e.g., inch, foot, yard, mile), mass (e.g., ounce, pound, ton), time (e.g., second, minute, hour, year), and capacity (fluid ounce, cup, pint, quart, gallon).

customary units [pg 93] The units of measure used in the customary measurement system (also the English system of measurement – e.g., inch, foot, yard, mile, ounce, cup, pint, quart, pound, ton).

Data [pg 105] Information collected and used to analyze a particular concept or situation.

denominator [pg 11, 14, 19, 28, 36-38, 40-41] The part of a fraction that is below the line and that functions as the divisor of the numerator.

diameter [pg 82, 84, 86-87] A chord of a circle which passes through the center of the circle.

decimal [pg 11, 12, 28, 92] Any any real number expressed in base 10.

digit [pg 11] Any of the Arabic numerals 1 to 9; one of the elements that combine to form numbers in a system other than the decimal system.

distance [pg 52,] The length of the line segment joining two points.

distributive property [pg 10] A property of real numbers that states that the product of the sum or difference of two numbers is the same as the sum or difference of their products.

Equation [pg 53] A mathematical sentence stating that two expressions are equal.

equivalent fractions [pg 12-13, 37] Two or more fractions that have the same quotient or that name the same region, part of a set, or part of a segment.

equivalent ratios [pg 19-20] Two ratios that are equal.

estimate [pg 112, 114] An answer that is an approximation based on rules of estimating.

evaluate [pg 51-52] To find the value of a mathematical expression.

event [pg 100] An outcome or group of outcomes to which a probability is assigned.

exponent [pg 42] A number that tells how many times the base is used as a factor; in an expression of the form b^a, a is called the exponent, b is the base, and b^a is a power of b.

exponential form [pg 42] A number written using exponents (e.g., b^a, x^3, y^2).

extremes [pg 20] The first and last terms in the ratios of a proportion.

Faces [pg 95] The sides of a rectangular prism.

formula [pg 52-53, 70] A mathematical statement, equation, or rule that shows a relationship between two or more quantities.

fraction [pg 11, 12, 28, 36-37, 40] A number that represents part of a whole, part of a set, or a quotient in the form a/b which can be read as a divided by b.

front-end estimation [pg 112-113] Using the leading, or left-most, digits to make an estimate quickly and easily.

Geometric figure [pg 70, 73] Any combination of points, lines, planes, or curves in two or three dimensions (e.g., square, circle, triangle, cube).

geometric shape [pg 70, 73] Any regular or irregular polygon, circle, or combination of geometric figures.

geometry [pg 60] The branch of mathematics that deals with the measurement, properties, and relationships of points, lines, angles, planes, and two- and three-dimensional figures.

graphically [pg 60, 103] To solve a problem or demonstrate a relationship using a number line or coordinate graph.

graphs [pg 103-104] Visual representations of data (e.g., bar, line).

Height [pg 70, 76] The perpendicular distance from a vertex to the line containing the opposite side of a plane figure; the length of a perpendicular from the vertex to the plane containing the base of a pyramid or cone; the length of a perpendicular between the planes containing the bases of a prism or cylinder.

hexagon [pg 62, 73] A polygon of six angles and six sides.

Identity element for addition [pg 10] The number in a set which when added to any number n in the set yields the given number.

identity element for multiplication [pg 10] The number in a set which when any number n in the set is multiplied by, yields the given number.

improper fraction [pg 36, 40] Fraction with a numerator larger than its denominator (e.g., $5/3$).

irregular polygon [pg 73-75] A polygon whose sides and angles are not all congruent.

LCD - least common denominator [pg 37-38] The least (or lowest) common multiple of two or more denominators.

length [pg 70, 76] The distance from one end of an object to the other end.

Mathematical expression [pg 50-51] A mathematical sentence whose truth value can be determined to be either true or false.

mean [pg 101-102] A measure of central tendency; the quotient obtained when the sum of the numbers in a set is divided by the number of addends.

means of a proportion [pg 20] The two middle terms in the ratios of a proportion.

measure [pg 92] To find the dimensions or quantity of an object or figure.

median [pg 101-102] The middle number of a set of numbers arranged in increasing or decreasing order; if there is no middle number, the median is the average of the two middle numbers.

metric system [pg 92] A system of measurement based on the decimal system; the standard unit of length is a meter, of capacity is a liter, and of mass is a gram.

metric units [pg 92] Units used in the metric system (e.g., kilo- = 1000, centi- $1/100$, milli- $1/1000$).

mixed number [pg 36, 38-41] A number composed of an integer and a proper fraction.

mode [pg 101-102] The number or members of a data set that occurs most frequently in the set.

multiplication [pg 9] A mathematical operation of combining groups of equal amounts; repeated addition; the inverse of division.

multiplicative inverse [pg 10] The reciprocal of a number; the number, which when multiplied by a given number, produces the multiplicative identity 1; in the set of real numbers the number a given number needs to be multiplied by to yield 1. (e.g., reciprocal of 5 is one fifth (1/5 or 0.2), the reciprocal of 0.25 is 1 divided by 0.25, or 4).

Natural numbers [pg 8] Either a positive integer (1, 2, 3, 4, ...) or a non-negative integer (0, 1, 2, 3, 4, ...); A set of counting numbers {1,2,3,4,...}.

negative number [pg 14] See absolute value. A number that is less than zero; located to the left of zero on the number line.

nonstandard measurement [pg 111, 115-116] The use of items as measurement tools that are not uniform in size.

nonstandard unit [pg 111, 115-116] Any tangible item that can be used to measure something.

number line [pg 14] A line on which each point represents a real number.

number system [pg 8] A system used to represent numbers.

numerator [pg 11, 14, 19, 28, 36-38, 40-41] The part of a fraction that is above the line and signifies the number to be divided by the denominator.

numeric expression [pg 50] Any combination of words, variables, constants, and/or operators that result in a number; also known as an arithmetic expression.

Operating symbols [pg 50] Used to translate word phrases into mathematical expressions (e.g., +, -, x, ÷).

operations [pg 36, 53] Procedures used to combine numbers, expressions, or polynomials into a single result.

order [pg 9, 13] To place numbers or objects in a sequential arrangement.

order of operations [pg 43, 51] A specified sequence in which mathematical operations are expected to be performed; an arithmetic expression is evaluated by following these ordered steps: (1) simplify within grouping symbols such as parentheses or brackets, starting with the innermost; (2) apply exponents – powers and roots; (3) perform all multiplications and divisions in order from left to right; (4) perform all additions and subtractions in order from left to right.

ordered pair [pg 60] A set of two numbers named in an order that matters; represented by (x, y) such that the first number, x, represents the x-coordinate and the second number, y, represents the y-coordinate when the ordered pair is graphed on the coordinate plane; each point on the coordinate plane has a unique ordered pair associated with it.

origin [pg 60] The point on the coordinate plane where the x- and y-axes intersect; has coordinates (0, 0).

Parallelogram [pg 62, 70, 76, 117] A quadrilateral with opposite sides parallel and equal.

pentagon [pg 62, 73] A polygon of five angles and five sides.

percent [pg 27-29, 31, 114] A number expressed in relation to 100 (symbol: %); also expressed as a fraction or a decimal.

perimeter [pg 63] The distance around a closed figure.

pi (π) [pg 84, 86-87] The ratio of the circumference of a circle to its diameter; pi is an irrational number with an approximate value of 3.14159 (symbol: π).

place value [pg 8-9] The value of a digit in a number based on its position (e.g., in the number 28, the 2 is in the tens place and the 8 is in the ones place).

point [pg 60, 63] An exact location in space; a point has no dimension.

polygon [pg 25, 62, 73, 76, 95] A closed plane figure formed by three or more line segments (E.G., square, circle, rectangle, triangle).

population [pg 99] A group of people, objects, or events that fit a particular description; in statistics, the set from which a sample of data is selected.

positive number [pg 14] Any number greater than zero or to the right of zero on the number line.

power [pg 42] The exponent of a number (e.g., b^3 means b to the third power or $b \times b \times b$).

precision of capacity [pg 95] Accuracy of measurement, generally the smaller the unit, the more precise the measurement.

predict [pg 100, 105] To be able to determine the next step or value (to make an educated guess), based on evidence or a pattern.

prism [pg 76, 95, 116] A polyhedron with two polygonal faces lying in parallel planes and with the other faces parallelograms.

probability [pg 100] The chance of an event occurring; the ratio of the number of favorable outcomes to the total number of possible outcomes; the probability of an event must be greater than or equal to 0 and less than or equal to 1.

product [pg 9] The number or expression resulting from the multiplication together of two or more numbers or expressions.

properties of real numbers [pg 9] Rules that apply to the operations with real numbers.

proportion [pg 20, 22, 29] An equation which states that two ratios are equivalent.

proportional reasoning [pg 20, 24, 29] Using the concept of proportions when analyzing and solving a mathematical situation.

Quadrant [pg 60] One of four sections of a coordinate grid separated by horizontal and vertical axes; they are numbered I, II, III, and IV, counterclockwise from the upper right.

quadrilateral [pg 62, 70, 73] A polygon with 4 sides and 4 angles.

Radius [pg 82, 84, 86-87] A line segment that extends from the center of a circle to any point on the circle.

range of a data set [pg 101-102] The difference between the greatest and the least values in a set of numbers.

rate [pg 18] A ratio that compares quantities of different units (e.g., miles per hour, price per pound, students per class, heartbeats per minute).

ratio [pg 18, 118] A comparison of two numbers or two like quantities by division.

rational number [pg 11, 14] Any number that can be expressed as a ratio in the form where a and b are integers and b divided by a non-zero integer.

real numbers [pg 10] A number that has no imaginary part (the set of all real numbers comprises the rationals and the irrationals).

reasonable estimate [pg 111,] An approximation of the result of a given problem or calculation using rational, logical procedures.

reciprocal of a fraction [pg 41] Either of a pair of numbers (as 2/3 and 3/2 or 9 and 1/9) whose product is one

rectangle [pg 62-63, 70, 116] A quadrilateral with four right angles.

rectangular prism [pg 95, 116] A three-dimensional figure (solid) that has two congruent and parallel faces that are polygons.

regular polygon [pg 73, 76, 95] A polygon in which all sides and all angles are congruent.

renaming a fraction [pg 39] Process of obtaining through multiplication or division of a fraction's numerator and denominator, two names for the same amount, called equivalent fractions (e.g., $1/2$ is the same as $2/4$).

repeating decimal [pg 11-12] A decimal in which one or more digits repeat infinitely.

rhombus [pg 62, 70] A parallelogram with two adjacent sides congruent (all four sides are congruent).

round a number [pg 84, 86, 113] To approximate the value of a whole number or decimal to a specific place value.

Sector of a circle [pg 86] The region of the circle formed by two radii and their intercepted arc.

similar triangles [pg 22-25] Triangles that have the same shape but not necessarily the same size; corresponding sides are in proportion and corresponding angles are congruent.

similar figures [pg 22, 25] Figures that have the same shape but not necessarily the same size (symbol: ~).

square [pg 62, 70, 117] A rectangle with two adjacent sides congruent (all four sides will be congruent).

standard form of a number [pg 8] A number is written in standard form when each digit is in a place value.

statistics [pg 100] The collection, organization, presentation, and analysis of data.

strategy [pg 112] A method or system of steps used to solve problems.

sum [pg 9] The whole amount as the result of adding numbers.

Terminating decimal [pg 11] A decimal whose digits do not repeat; all terminating decimals are rational numbers.

trapezoid [pg 62, 70, 73, 117-118] A quadrilateral with exactly one pair of parallel sides.

tree diagram [pg 100] a method to display and count possible outcomes in probability problems.

triangle [pg 62, 70, 73, 117] A polygon with three sides and three angles.

Variable [pg 49-51, 53] A symbol used to represent a number or group of numbers in an expression or an equation.

verbal expression [pg 50] A phrase stating a relationship; can be translated into a mathematical/algebraic expression.

verbal form [pg 50] A mathematical expression or relationship using words rather than symbols.

vertex [pg 63] (1) The common endpoint of two sides of a polygon; (2) the common endpoint of two rays that form an angle; (3) the common point where two or more edges of a three-dimensional solid meet.

volume [pg 76, 95, 116] The number of cubic units needed to fill a solid figure.

Whole numbers [pg 8, 36] The set of counting numbers plus zero; {0,1,2,3, ...}.

width [pg 70, 76] One dimension of a two- or three-dimensional figure.

written symbol [pg 50] A sign used to represent something such as an operation (+, −, x , ÷), a relationship (=, ≠, <, >, ≤, ≥), or a special quantity.

X-axis [pg 60] The horizontal axis; the line whose equation is $y = 0$.

x-coordinate [pg 60, 63] identifies how far up or down from the origin the point is located.

Y-axis [pg 60] The vertical axis; the line whose equation is $x = 0$.

y-coordinate [pg 60, 63] Identifies how far to the left or right of the origin the point is located.

Zero property of addition [pg 10] The property that states that the sum of a number and zero is that same number (i.e., $a + 0 = a$ for all a).

zero property of multiplication [pg 10] The property that states that the product of any number and zero is always zero (e.g., $a \cdot 0 = 0$).

CORRELATION TO CURRICULUM OUTLINE

<u>Lesson One</u>
<u>Number Systems</u>   . . . . . .Student Review **p 7**

1.1 - Whole Numbers
6.N.1 – Read and write whole numbers to trillions

1.2 - Properties for Addition and Multiplication
6.N.2 – Define and identify the commutative and associative properties of addition and multiplication
6.N.3 – Define and identify the distributive property of multiplication over addition
6.N.4 – Define and identify the identity and inverse properties of addition and multiplication
6.N.5 – Define and identify the zero property of multiplication

1.3 - Rational Numbers
6.N.13 – Define absolute value of a rational number
6.N.15 – Order rational numbers
6.N.20 – Represent fractions as terminating or repeating decimals
6.N.21 – Find multiple representations of rational numbers

1.4 - Rational Numbers on a Number Line
6.N.14 – Locate rational numbers on a number line

<u>Lesson Two – Ratios, Rates,</u>
<u>Proportions, and Percents</u> . . . .Student Review **p 17**

2.1 - Ratios and Rates
6.N.6 – Understand the concept of rate
6.N.8 – Distinguish the difference between rate and ratio

2.2 - Equivalent Ratios
6.N.7 – Express equivalent ratios as a proportion

2.3 - Proportions
6.N.9 – Solve proportions using equivalent fractions
6.N.10 – Verify the proportionality using the product of the means equals the product of the extremes

2.4 - Similar Figures
6.G.1 – Calculate the length of corresponding sides of similar triangles, using proportional reasoning

2.5 - Percents
6.N.11 – Read, write, and identify percents of a whole (0% to 100%)

2.6 – Changing Percents to Fractions, and Decimals

2.7 – Calculating Percents
6.N.12 – Solve percent problems involving percent, rate, and base

<u>Lesson Three –</u>
<u>Operations</u>   . . . . . .Student Review **p 35**

3.1 - Fractions and Mixed Numbers

3.2 - Add and Subtract Fractions and Mixed Numbers
6.N.16 – Add and subtract fractions with unlike denominators
6.N.18 – Add, subtract, multiply and divide mixed numbers with unlike denominators

3.3 - Multiply and Divide Fractions and Mixed Numbers
6.N.17 – Multiply and divide fractions with unlike denominators
6.N.18 – Add, subtract, multiply and divide mixed numbers with unlike denominators
6.N.19 – Identify the multiplicative inverse of a number

8.2 - Customary Units of Capacity
6.M.2 – Identify customary units of capacity
6.M.3 – Identify equivalent customary units of capacity

8.3 - Measurement and Precision of Capacity
6.M.6 – Determine the tool and technique to measure with an appropriate level of precision: capacity

8.4 - Volume of a Rectangular Prism
6.M.1 – Measure capacity and calculate volume of a rectangular prism

9.1 - Probability
5.S.5 – List the possible outcomes for a single-event experiment
5.S.6 – Record experiment results using fractions / ratios
5.S.7 – Create a sample space and determine the probability of a single even, given a simple experiment

9.2 - Mean, Median, Mode, and Range
6.S.5 – Determine the mean, mode, and median for a given set of data
6.S.6 – Determine the range for a given set of data

9.3 - Reading and Interpreting Graphs
6.S.7 – Read and interpret graphs

9.4 - Predictions Made from Data
6.S.8 – Justify predictions made from data

10.1 – Estimation Strategies
6.N.27 – Justify the reasonableness of answers using estimation

10.2 – Estimating Percents
6.N.26 – Estimate a percent of quantity (0% to 100%)

10.3 - Measurement Estimation
6.M.8 – Justify the reasonableness of estimates
6.M.9 – Determine the personal references for capacity

10.4 – Estimate Volume, Area and Circumference
6.M.7 – Estimate volume, area, and circumference

Notes:

Notes:

Notes:

Notes: